U.S. Department of Transportation
National Highway Traffic Safety Administration

DOT HS 811 665   August 2012

Relationships Between Fatality Risk, Mass, and Footprint in Model Year 2000-2007 Passenger Cars and LTVs

Final Report

This publication is distributed by the U.S. Department of Transportation, National Highway Traffic Safety Administration, in the interest of information exchange. The opinions, findings, and conclusions expressed in this publication are those of the authors and not necessarily those of the Department of Transportation or the National Highway Traffic Safety Administration. The United States Government assumes no liability for its content or use thereof. If trade or manufacturers' names or products are mentioned, it is because they are considered essential to the object of the publication and should not be construed as an endorsement. The United States Government does not endorse products or manufacturers.

Suggested APA Format Reference:

Kahane, C. J. (2012, August). *Relationships Between Fatality Risk, Mass, and Footprint in Model Year 2000-2007 Passenger Cars and LTVs – Final Report*. (Report No. DOT HS 811 665). Washington, DC: National Highway Traffic Safety Administration.

Technical Report Documentation Page

1. Report No. DOT HS 811 665	2. Government Accession No.	3. Recipient's Catalog No.
4. Title and Subtitle Relationships Between Fatality Risk, Mass, and Footprint in Model Year 2000-2007 Passenger Cars and LTVs – Final Report		5. Report Date August 2012
		6. Performing Organization Code
7. Author(s) Charles J. Kahane, Ph.D.		8. Performing Organization Report No.
9. Performing Organization Name and Address Office of Vehicle Safety National Highway Traffic Safety Administration Washington, DC 20590		10. Work Unit No. (TRAIS)
		11. Contract or Grant No.
12. Sponsoring Agency Name and Address National Highway Traffic Safety Administration 1200 New Jersey Avenue SE. Washington, DC 20590		13. Type of Report and Period Covered NHTSA Technical Report
		14. Sponsoring Agency Code

15. Supplementary Notes

16. Abstract

Mass reduction while holding a vehicle's footprint (size) constant is a potential strategy for meeting footprint-based CAFE and GHG standards. An important corollary issue is the possible effect of mass reduction that maintains footprint on fatal crashes. One way to estimate these effects is statistical analyses of societal fatality rates per VMT, by vehicles' mass and footprint, for the current on-road vehicle fleet. Societal fatality rates include occupants of all vehicles in the crash as well as pedestrians. The analyses comprised MY 2000-2007 cars and LTVs in CY 2002-2008 crashes. Fatality rates were derived from FARS data, 13 State crash files, and registration and mileage data from R.L. Polk. The table presents the estimated percent increase in societal fatality rates per 100-pound mass reduction while holding footprint constant for five classes of vehicles:

MY 2000-2007 CY 2002-2008	Fatality Increase (%) Per 100-Pound Mass Reduction While Holding Footprint Constant	
	Point Estimate	95% Confidence Bounds
Cars < 3,106 pounds	1.56	+ .39 to +2.73
Cars ≥ 3,106 pounds	.51	- .59 to +1.60
CUVs and minivans	- .37	-1.55 to + .81
Truck-based LTVs < 4,594 pounds	.52	- .45 to +1.48
Truck-based LTVs ≥ 4,594 pounds	- .34	- .97 to + .30

Only the 1.56 percent risk increase in the lighter cars is statistically significant. There are non-significant increases in the heavier cars and the lighter truck-based LTVs and non-significant societal benefits for mass reduction in CUVs, minivans, and the heavier truck-based LTVs. Based on these results, potential combinations of mass reductions that maintain footprint and are proportionately somewhat higher for the heavier vehicles may be safety-neutral or better as point estimates and, in any case, unlikely to significantly increase fatalities. The primarily non-significant results are not due to a paucity of data, but because the societal effect of mass reduction while maintaining footprint, if any, is small.

17. Key Words		18. Distribution Statement Document is available to the public from the National Technical Information Service www.ntis.gov.	
19. Security Classif. (Of this report) Unclassified	20. Security Classif. (Of this page) Unclassified	21. No. of Pages 189	22. Price

Form DOT F 1700.7 (8-72)   Reproduction of completed page authorized

Table of Contents

List of abbreviations iv

Executive summary vii
    A new analysis of fatality risk, mass, and footprint vii
    What's new in MY 2000-2007 vehicles and in this report? ix
    Results xii
    Sensitivity tests xiv
    Scope and limitations of the analyses xvi

1. Background for a new analysis of fatality risk, mass, and footprint 1
    1.1   NHTSA's 2010 report 1
    1.2   Next steps proposed in the 2010 final rule for 2012-2016 CAFE 3
    1.3   NHTSA's November 2011 preliminary report 5
    1.4   Developments in MY 2000-2007 vehicles 5
    1.5   Earlier studies 8
    1.6   Hypothetical relationships between mass and societal fatality risk in the data 9
    1.7   Culpability in 2-vehicle crashes: relationships with mass and footprint 12
    1.8   Peer-review recommendations on NHTSA's 2010 report 18
    1.9   Limitations of statistical analyses of historical crash and exposure data 19

2. New databases to study fatalities per billion VMT 21
    2.0   Summary 21
    2.1   Vehicle classification, curb weight, footprint, and other attributes 22
    2.2   Fatal crash involvements: FARS data reduction 25
    2.3   Vehicle registration years: Polk NVPP data reduction 29
    2.4   Annual VMT: odometer readings from Polk 29
    2.5   Induced-exposure crashes: State data reduction 31
    2.6   Assembling the analysis data files 34

3. Results 41
    3.0   Summary 41
    3.1   Regression setup 41
    3.2   Three regression examples 45
    3.3   Mass and footprint coefficients for the 27 basic regressions 51
    3.4   The effect of mass within deciles of footprint 55
    3.5   Overall effect of mass reduction by vehicle class and its confidence bounds 60
    3.6   Combined annual effect of mass reduction in several classes of vehicles 70
    3.7   Effect of reducing mass and footprint (downsizing) 75
    3.8   Regression analyses including IIHS crash-test ratings 77
    3.9   Overall effect by vehicle class can change if the on-road fleet changes mass 82
    3.10 Effect of mass reduction for drivers with BAC < .08 or with BAC = 0 84

4.	Sensitivity tests and response to comments on the preliminary report	89
	4.0 Summary	89
	4.1 Comments on NHTSA's 2011 preliminary report	89
	4.2 Sensitivity test results	91
	4.3 Discussion of individual sensitivity tests	96
	4.4 Effect on the case vehicle's versus the other vehicle's occupants	110
	4.5 Indexing curb weight to footprint and other ways to address multicollinearity	114
	4.6 Response to other comments	118

References	122
Appendix A: MY 2000-2007 SUVs considered crossover utility vehicles in this report	127
Appendix B: Codebook for the database of fatal crash involvements	129
Appendix C: Codebook for the database of induced-exposure crash involvements	139
Appendix D: Volpe and regression coefficients for sensitivity tests	145

List of Abbreviations

ABS	Antilock brake system
AWD	All-wheel drive
BAC	Blood alcohol concentration, measured in grams per deciliter (g/dL)
BMW	Bayerische Motoren Werke
CAFE	Corporate Average Fuel Economy
CARB	California Environmental Protection Agency Air Resources Board
cg	Center of gravity
CUV	Crossover utility vehicle
CY	Calendar year
DC, D.C.	District of Columbia
df	Degrees of freedom
DOE	United States Department of Energy
DOT	United States Department of Transportation
DRI	Dynamic Research, Inc.
DWI	Driving while intoxicated
EISA	Energy Independence and Security Act of 2007
EMS	Emergency medical services
EPA	United States Environmental Protection Agency
ESC	Electronic stability control
FARS	Fatality Analysis Reporting System, a census of fatal crashes in the United States since 1975
FIPS	Federal Information Processing Standard
FMVSS	Federal Motor Vehicle Safety Standard
FPC	Finite population correction

FRIA	Final Regulatory Impact Analysis
FWD	Front-wheel drive
GES	General Estimates System of NASS
GHG	Greenhouse gases
GMC	General Motors Corporation
GVWR	Gross vehicle weight rating, specified by the manufacturer, equals the vehicle's curb weight plus maximum recommended loading
HLDI	Highway Loss Data Institute
ICCT	International Council on Clean Transportation
IIHS	Insurance Institute for Highway Safety
LBNL	Lawrence Berkeley National Laboratory, Berkeley, CA
LTV	Light trucks and vans, includes pickup trucks, SUVs, minivans, and full-size vans
mpg	Miles per gallon
MY	Model year
NAS	National Academy of Sciences
NASS	National Automotive Sampling System, a probability sample of police-reported crashes in the United States since 1979, investigated in detail
NCAP	New Car Assessment Program
NHTS	National Household Transportation Survey
NHTSA	National Highway Traffic Safety Administration
NPRM	Notice of Proposed Rulemaking
NVPP	R.L. Polk's National Vehicle Population Profiles
OMB	Office of Management and Budget of the United States Government
PRIA	Preliminary Regulatory Impact Analysis
PSU	Primary sampling unit

RWD	Rear-wheel drive
SAS®	Statistical and database management software produced by SAS Institute, Inc.
SRS	Simple random sample, simple random sampling
SUV	Sport utility vehicle
UMTRI	University of Michigan Transportation Research Institute
VIF	Variance inflation factor
VIN	Vehicle Identification Number
VMT	Vehicle miles of travel

Executive Summary

A new analysis of fatality risk, mass, and footprint

On May 7, 2010, the National Highway Traffic Safety Administration (NHTSA) and the Environmental Protection Agency (EPA) published a joint final rule to establish Corporate Average Fuel Economy (CAFE) standards and greenhouse-gas (GHG) emission standards for passenger cars and light trucks manufactured in model years (MY) 2012-2016.[1] The standards for MY 2012-2016 are "footprint-based," with footprint being defined as a measure of a vehicle's size, roughly equal to the wheelbase times the average of the front and rear track widths. Basing standards on vehicle footprint ideally helps to discourage vehicle manufacturers from downsizing their vehicles, because the agencies set higher (more stringent) mpg targets for smaller-footprint vehicles, but would not similarly discourage mass reduction that maintains footprint while potentially improving fuel economy. Several technologies, such as substitution of light, high-strength materials for conventional materials during vehicle redesigns, have the potential to reduce weight and conserve fuel while maintaining a vehicle's footprint and maintaining or possibly improving the vehicle's structural strength and handling.

On November 16, 2011, NHTSA and EPA published a joint Notice of Proposed Rulemaking (NPRM) to establish CAFE and GHG standards for passenger cars and light trucks manufactured in model years (MY) 2017-2025.[2] The proposed standards for MY 2017-2025 are again footprint-based.

In considering what technologies are available for improving fuel economy, including mass reduction, an important corollary issue for NHTSA to consider is the potential effect that those technologies may have on safety. NHTSA has thus far specifically considered the likely effect of mass reduction that maintains footprint on fatal crashes. The relationship between a vehicle's mass, size, and fatality risk is complex, and it varies in different types of crashes. NHTSA, along with others, has been examining this relationship for over a decade. The safety chapter of NHTSA's April 2010 final regulatory impact analysis (FRIA) of CAFE standards for MY 2012-2016 passenger cars and light trucks included a statistical analysis of relationships between fatality risk, mass, and footprint in MY 1991-1999 passenger cars and LTVs (light trucks and vans), based on calendar year (CY) 1995-2000 crash and vehicle-registration data.[3]

The principal findings and conclusions of NHTSA's 2010 report were that mass reduction in the lighter cars, even while holding footprint constant would significantly increase fatality risk, whereas mass reduction in the heavier LTVs would significantly reduce societal fatality risk, because it would reduce the fatality risk of occupants of lighter vehicles colliding with those heavier LTVs. NHTSA concluded that, as a result, any *reasonable* combination of mass reductions that held footprint constant in MY 2012-2016 vehicles – concentrated, at least to

---

[1] 75 Fed. Reg. 25324 (May 7, 2010).
[2] 76 Fed. Reg. 74854 (December 1, 2011).
[3] Kahane, C. J. (2010). "Relationships Between Fatality Risk, Mass, and Footprint in Model Year 1991-1999 and Other Passenger Cars and LTVs," *Final Regulatory Impact Analysis: Corporate Average Fuel Economy for MY 2012-MY 2016 Passenger Cars and Light Trucks*. Washington, DC: National Highway Traffic Safety Administration, pp. 464-542, available at http://www.nhtsa.gov/staticfiles/rulemaking/pdf/cafe/CAFE_2012-2016_FRIA_04012010.pdf.

some extent, in the heavier LTVs and limited in the lighter cars – would likely be approximately safety-neutral; it would not significantly increase fatalities and might well decrease them.

NHTSA's 2010 report partially agreed and partially disagreed with analyses published during 2003-2005 by Dynamic Research, Inc. (DRI). NHTSA and DRI both found a significant protective effect for footprint and that reducing mass and footprint together (downsizing) on smaller vehicles was harmful. On the other hand, DRI's analyses estimated significant overall benefits for mass reduction in all passenger cars and LTVs if wheelbase and track width were maintained, whereas NHTSA's report showed an overall benefit only in the heavier LTVs, but for other classes of vehicles, benefits only in some types of crashes. Much of NHTSA's 2010 report as well as recent work by DRI involved sensitivity tests on the databases and models and generated a range of estimates somewhere between the initial DRI and NHTSA results.[4]

Immediately after issuing the final rule for MYs 2012-2016 CAFE and GHG standards in May 2010, NHTSA and EPA began work on the next joint rulemaking to develop CAFE and GHG standards for MYs 2017 and beyond, issuing a Notice of Intent to conduct a joint rulemaking in October 2010 with an accompanying Interim Technical Assessment Report, a Supplemental Notice of Intent in December 2010, and the NPRM in November 2011.[5] The preamble to the 2012-2016 final rule stated that NHTSA, working closely with EPA and the Department of Energy (DOE), would perform a new statistical analysis of the relationships between fatality rates, mass and footprint, updating the crash and exposure databases to the latest available model years and refreshing the methodology in response to peer reviews of the 2010 report and taking into account changes in vehicle technologies. The previous databases of MY 1991-1999 vehicles in CY 1995-2000 crashes have become outdated as new safety technologies, vehicle designs and materials were introduced. The new databases comprising MY 2000-2007 vehicles in CY 2002-2008 crashes are the most up-to-date possible, given the processing time for crash data and the need for enough crash cases to permit statistically meaningful analyses. NHTSA has made the new databases available to the public at http://www.nhtsa.gov/fuel-economy, enabling other researchers to analyze the same data and hopefully minimizing discrepancies in the results that would have been due to inconsistencies across databases.[6]

NHTSA published its preliminary analysis of the new MY 2000-2007 databases in November 2011, simultaneous with the NPRM for MY 2017-2025 CAFE.[7] The preliminary report was

---

[4] Van Auken, R. M., and Zellner, J. W. (2003). *A Further Assessment of the Effects of Vehicle Weight and Size Parameters on Fatality Risk in Model Year 1985-98 Passenger Cars and 1986-97 Light Trucks.* Report No. DRI-TR-03-01. Torrance, CA: Dynamic Research, Inc.; Van Auken, R. M., and Zellner, J. W. (2005a). *An Assessment of the Effects of Vehicle Weight and Size on Fatality Risk in 1985 to 1998 Model Year Passenger Cars and 1985 to 1997 Model Year Light Trucks and Vans.* Paper No. 2005-01-1354. Warrendale, PA: Society of Automotive Engineers; Van Auken, R. M., and Zellner, J. W. (2005b). *Supplemental Results on the Independent Effects of Curb Weight, Wheelbase, and Track on Fatality Risk in 1985-1998 Model Year Passenger Cars and 1986-97 Model Year LTVs.* Report No. DRI-TR-05-01. Torrance, CA: Dynamic Research, Inc.; Van Auken, R.M., and Zellner, J. W. (2012a). *Updated Analysis of the Effects of Passenger Vehicle Size and Weight on Safety, Phase I.* Report No. DRI-TR-11-01. (Docket No. NHTSA-2010-0152-0030). Torrance, CA: Dynamic Research, Inc.
[5] All documents are available at http://www.nhtsa.gov/fuel-economy. The Technical Assessment Report did not contain an assessment of the safety effects of potential standards, deferring that until the NPRM.
[6] 75 Fed. Reg. 25324 (May 7, 2010); the discussion of planned statistical analyses is on pp. 25395-25396.
[7] Kahane, C. J. (2011). *Relationships Between Fatality Risk, Mass, and Footprint in Model Year 2000-2007 Passenger Cars and LTVs – Preliminary Report.* (Docket No. NHTSA-2010-0152-0023). Washington, DC: National Highway Traffic Safety Administration.

issued for public comment[8] and peer-reviewed according to the Office of Management and Budget's (OMB) guidelines. Public and peer reviews were completed by February 2012. NHTSA then slightly revised the MY 2000-2007 databases by adding CY 2008 crash data that recently had become available from certain States and updating the estimates of annual VMT. NHTSA has again made the latest databases available to the public at http://www.nhtsa.gov/fuel-economy. This final report updates the preliminary report's analyses with the slightly revised MY 2000-2007 databases and adds a new, fourth chapter of analyses in response to the comments and reviews, primarily sensitivity tests.

What's new in MY 2000-2007 vehicles and in this report?

The most noticeable change in MY 2000-2007 vehicles from MY 1991-1999 has been the increase in crossover utility vehicles (CUV), which are SUVs of unibody construction, often but not always built upon a platform shared with passenger cars. CUVs have blurred the distinction between cars and trucks. The analyses of MY 2000-2007 data – in this report as well as NHTSA's 2011 preliminary report – treat CUVs and minivans as a separate vehicle class, because they differ in some respects from pickup-truck-based LTVs and in other respects from passenger cars. In NHTSA's 2010 report based on the MY 1991-1999 databases, the many different types of LTVs were combined in a single analysis and NHTSA believes that this may have made the analyses too complex and might have contributed to some of the uncertainty in the results.

MY 2000-2007 vehicles of all types are also heavier and larger than their MY 1991-1999 counterparts. The average mass of passenger cars increased by 5 percent from 2000 to 2007. The mass of the average pickup truck increased by 19 percent. Other types of vehicles became heavier, on the average, by intermediate amounts. Across the fleet, the increase in footprint ranged from 3 percent in cars to 10 percent in pickup trucks. There are several reasons for these increases: during this time frame, some of the lighter make-models were discontinued; many models were redesigned to be heavier and larger; and consumers more often selected stretched versions such as crew cabs in their new-vehicle purchases.

MY 2000-2007 was also a time of great safety improvement. Frontal air bags were standard on all new vehicles; electronic stability control (ESC) and curtain air bags advanced from rare options to half of the market; many more truck-based LTVs met voluntary compatibility standards. ESC is a technology that will change the relative distribution of crash types by preventing a substantial proportion of the rollovers and impacts with fixed objects. Ratings on the Insurance Institute for Highway Safety's (IIHS) frontal-offset test improved dramatically, from 24 percent rated "good" overall in 2000 to 86 percent in 2007.[9] Additionally, during MY 2000-2007 (and for some years before), many of the lightest and smallest vehicles were phased out in response to dwindling sales and changing consumer preferences. Many particularly poor safety performers, which were often among the lightest and smallest vehicles, were phased out. To account for these changes in the on-road fleet, the new analyses include variables on the availability of ESC, curtain and side air bags, compatibility certification, and IIHS ratings.

---

[8] 76 Fed. Reg. 73008 (November 28, 2011).
[9] Ratings for individual makes and models may be found at www.iihs.org.

Revisions to the analyses: The basic analysis method is unchanged from the 2011 preliminary report and is basically the same as in NHTSA's 2010 report: cross-sectional analyses of the societal fatality rate per billion vehicle miles of travel (VMT) by mass and footprint, while controlling for driver age, gender, and other factors, in separate logistic regressions by vehicle class and crash type. "Societal" fatality rates include fatalities to occupants of all the vehicles involved in the collisions, plus any pedestrians. The data is now MY 2000-2007 vehicles in CY 2002-2008, updated from the previous database of MY 1991-1999 vehicles in CY 1995-2000. The new data has accurate VMT estimates, derived in part from a file of odometer readings by make, model, and model year recently developed by R.L. Polk and purchased by NHTSA.[10] The vehicles are now grouped into three classes rather than two, for the reasons discussed above: passenger cars (including both 2-door and 4-door cars); CUVs and minivans; and truck-based LTVs.

There are also nine types of crashes rather than the six in the 2010 report; specifically, "collision with car" and "collision with LTV" in the previous analysis have been replaced by four types of crashes: collision with car, CUV, or minivan < 3,082 pounds; collision with car, CUV, or minivan ≥ 3,082 pounds; collision with truck-based LTV < 4,150 pounds; and collision with truck-based LTV ≥ 4,150 pounds. Splitting the "other" vehicles into a lighter and a heavier group permits more accurate analyses of the mass effect in collisions of two light vehicles. Grouping partner-vehicle CUVs and minivans with cars rather than LTVs is more appropriate because their front-end profile and rigidity more closely resembles a car than a typical truck-based LTV.

The curb weight of passenger cars is formulated, as in the 2010 report, as a two-piece linear variable in order to estimate one effect of mass reduction in the lighter cars and another effect in the heavier cars. The boundary between "lighter" and "heavier" cars is 3,106 pounds (which is the median mass of MY 2000-2007 cars in fatal crashes, up from 2,950 in 1991-1999). Likewise, for truck-based LTVs, curb weight is a two-piece linear variable with the boundary at 4,594 pounds (again, the 2000-2007 median, much higher than the median of 3,870 in 1991-1999). Curb weight is formulated as a simple linear variable for CUVs and minivans: because CUVs and minivans account for a much smaller share of new-vehicle sales, there is much less crash data available than for cars or truck-based LTVs.

For a given vehicle class and weight range (if applicable), the regression coefficients for mass (while holding footprint constant) in the nine types of crashes are averaged, weighted by the number of baseline fatalities that would have occurred for the subgroup MY 2004-2007 vehicles in CY 2004-2008 if these vehicles had all been equipped with ESC. The adjustment for ESC, a new feature of the analysis, takes into account that the results will be used to analyze effects of mass reduction in future vehicles, which will all be ESC-equipped, as required by NHTSA's regulations. A similar adjustment to the baseline fatalities probably should have been applied in the 2010 report, but was not.

Techniques have been added to test significance and to estimate 95% confidence bounds (sampling error) for each mass effect and to estimate the combined annual effect of removing

---

[10] In the 1991-1999 database, VMT was estimated only by vehicle class, based on NASS CDS data.

100 pounds of mass from every vehicle (or of removing different amounts of mass from the various classes of vehicles), while holding footprint constant.

NHTSA considered the near multicollinearity of mass and footprint to be a major issue in the 2010 report and voiced concern about inaccurately estimated regression coefficients.[11] The high correlations between mass and footprint and variance inflation factors (VIF) have not changed from MY 1991-1999 to MY 2000-2007; large vehicles continued to be, on the average, heavier than small vehicles to the same extent as in the previous decade.[12] Nevertheless, multicollinearity appears to be less of a problem in the analysis this time. The "decile" analysis comparing fatality rates of vehicles of different mass but nearly identical footprint (modified in response to peer-review comments to control for factors such as driver age and gender) largely corroborates the main regression results. Whereas perhaps 4 of the 27 basic regressions still display possible symptoms of near multicollinearity, namely exceptionally strong coefficients in opposite directions for mass and footprint, the positive coefficient goes twice to mass, twice to footprint: in short, there appears to be no systematic bias. Separating the CUVs and minivans from the other LTVs may also have helped to stabilize the results. NHTSA has no other explanation of why multicollinearity became less of a problem, except this: when there are only a few (2-4) regressions in each report that seem to display symptoms of multicollinearity, it could readily happen by chance that all of them give the positive coefficient to the same variable in one report (curb weight in 2010) but split close to 50-50 this time.

Another issue noted in the 2010 report was the historical trend of lighter and smaller vehicles to be less well driven – as evidenced, for example, in the higher odds of culpability for their drivers, even after controlling for the driver's age and gender and the vehicle's safety technologies. The trend contributes to the higher fatality risk of the lighter and smaller vehicles in statistical analyses. It is unknown if a vehicle's lightness or smallness in any way contributes to how people drive it or if the trend merely reflects that better drivers, on the average, prefer larger vehicles. The trend is still there in the new data, but it appears to have diffused. In the earlier database, the trend was attributed primarily to mass, not footprint. Now it is about equally attributed to mass and footprint.

In the 2010 report, largely because of those two issues – multicollinearity and the trend of lighter/smaller vehicles to be less well driven – NHTSA supplemented the actual regression results with alternative "upper-estimate" and "lower-estimate" scenarios that set aside some of the individual regression coefficients and replaced them with a range of estimates derived from other sources. Because these issues no longer seem critical, this report presents only a single set of estimates based on the actual regression results and it does not include such alternative scenarios.

---

[11] Van Auken and Green also discussed the issue in their presentations at the NHTSA Workshop on Vehicle Mass-Size-Safety in Washington, DC on February 25, 2011, http://www.nhtsa.gov/Laws+&+Regulations/CAFE+-+Fuel+Economy/NHTSA+Workshop+on+Vehicle+Mass-Size-Safety.

[12] Greene, W. H. (1993). *Econometric Analysis*, Second Edition. New York: Macmillan Publishing Company, pp. 266-268; Allison, P.D. (1999), *Logistic Regression Using the SAS System*. Cary, NC: SAS Institute Inc., pp. 48-51. VIF scores are in the 6-9 range for curb weight and footprint in NHTSA's new database – i.e., in the somewhat unfavorable 2.5-10 range where near multicollinearity begins to become a concern in logistic regression analyses.

Results

The immediate purpose of this report is to develop five parameters that the CAFE Compliance and Effects Modeling System (usually referred to as the "Volpe model," developed for NHTSA by the Volpe National Transportation Systems Center) will use in the FRIA to estimate the safety effects, if any, of the modeled mass reductions in MY 2017-2025 vehicles over their lifetime. The five numbers are the overall percentage increases or decreases, per 100-pound mass reduction while holding footprint constant, in societal fatalities involving five classes of vehicles/weight ranges:

Principal Findings: MY 2000-2007, CY 2002-2008
Fatality Increase (%) per 100-Pound Mass Reduction While Holding Footprint Constant

	Point Estimate	95% Confidence Bounds
Cars < 3,106 pounds	1.56	+ .39 to +2.73
Cars ≥ 3,106 pounds	.51	- .59 to +1.60
CUVs and minivans	-.37	-1.55 to + .81
Truck-based LTVs < 4,594 pounds	.52	- .45 to +1.48
Truck-based LTVs ≥ 4,594 pounds	-.34	- .97 to + .30

As discussed in more detail below, this analysis finds that societal fatality risk increases by 1.56 percent if mass is reduced by 100 pounds in the lighter cars. This is the only statistically significant effect found in the current analysis; it is the only one with confidence bounds that exclude zero. (The confidence bounds estimate only the sampling error internal to the data used in the specific analysis that generated the point estimate. Additional uncertainty, more difficult to quantify, could be attributed to sensitivity of the point estimate to modifying features of the analysis – e.g., selection of control variables.) There are non-significant increases in societal fatality risk for mass reduction in the heavier cars and the lighter truck-based LTVs. There are non-significant societal benefits for mass reduction in CUVs, minivans, and the heavier truck-based LTVs.

The five numbers have barely changed from NHTSA's November 2011 preliminary report. The changes in the point estimates are quite small relative to the sampling-error confidence bounds of each estimate:

November 2011 Preliminary Report: MY 2000-2007, CY 2002-2008
Fatality Increase (%) per 100-Pound Mass Reduction While Holding Footprint Constant

	Point Estimate	*95% Confidence Bounds*
Cars < 3,106 pounds	*1.44*	*+ .29 to +2.59*
Cars ≥ 3,106 pounds	*.47*	*- .58 to +1.52*
CUVs and minivans	*-.46*	*-1.75 to + .83*
Truck-based LTVs < 4,594 pounds	*.52*	*- .43 to +1.46*
Truck-based LTVs ≥ 4,594 pounds	*-.39*	*-1.06 to + .27*

It is interesting to compare the new results to NHTSA's 2010 analysis of MY 1991-1999 vehicles in CY 1995-2000, especially the new point estimate to the "actual regression result scenario" in the 2010 report:

2010 Report: MY 1991-1999, CY 1995-2000
Fatality Increase (%) per 100-Pound Mass Reduction While Holding Footprint Constant

	Actual Regression Result Scenario	Upper-Estimate Scenario	Lower-Estimate Scenario
Cars < 2,950 pounds	2.21	2.21	1.02
Cars ≥ 2,950 pounds	0.90	0.90	0.44
LTVs < 3,870 pounds	0.17	0.55	0.41
LTVs ≥ 3,870 pounds	-1.90	-0.62	-0.73

The new results are directionally the same as in 2010: fatality increase in the lighter cars, safety benefit in the heavier LTVs. But the effects may have become weaker at both ends. (The agency does not consider this conclusion to be definitive because of the relatively wide confidence bounds of the estimates.) The fatality increase in the lighter cars tapered off from 2.21 percent to 1.56 percent while the societal benefit of mass reduction in the heaviest LTVs diminished from 1.90 percent to 0.34 percent and is no longer statistically significant.

NHTSA believes that the changes may be due to a combination of "real" factors (characteristics of the newer vehicles) and revisions to the analysis. Above all, many cars with poor safety performance, which were often light, small cars, were discontinued by 2000 or during 2000-2007. The tendency of light, small vehicles to be driven poorly, while still there, is not as strong as it used to be – perhaps in part because safety improvements in lighter and smaller vehicles have made some good drivers more willing to buy them. At the other end of the spectrum, blocker beams and other voluntary compatibility improvements in LTVs as well as compatibility-related self-protection improvements to cars have made the heavier LTVs somewhat less aggressive in collisions with lighter vehicles (although the effect of mass disparity remains). This report's analysis of CUVs and minivans as a separate class of vehicles may have relieved some inaccuracies in the 2010 regression results for LTVs. Interestingly, the new actual-regression results are quite close to the previous report's "lower-estimate scenario," which was an attempt to adjust for supposed inaccuracies in some regressions and for a seemingly excessive trend toward higher crash rates in smaller and lighter cars.

Will the effects continue to diminish in the future? NHTSA believes it would be risky and inadvisable to extrapolate from only two data points. Continued safety improvement in the lighter and smaller vehicles, with increasing consumer awareness of their safety could potentially point in that direction. On the other hand, the 2000-2007 period may have been unique in that there were few really small vehicles – i.e., there were relatively few data points to influence the regression results at the lowest end of the mass spectrum where, perhaps, the relationships of mass and footprint to safety might be even stronger than in the medium-small vehicles. A new generation of substantially smaller and lighter vehicles, even if designed to a high level of safety, might influence future regression results simply by providing more data points at the low end of the spectrum.

One way of looking at the estimated safety effects, albeit one that is not truly under consideration, would be an across-the-board 100-pound reduction in all vehicles. The estimated effect of simultaneously reducing all new vehicles by exactly 100 pounds (a higher proportion of the mass of a light car than of a heavy LTV) while maintaining footprint is that fatalities would increase by 157 ± 196 per year – not statistically significant, but again, also not really under consideration. A second combination that would still average over the fleet to 100 pounds per vehicle – namely, a proportionate 2.57 percent reduction in the mass of all vehicles while maintaining footprint – would increase fatalities by an estimated 108 ± 196 lives per year, again not statistically significant. A third combination that would still average over the fleet to 100 pounds per vehicle – namely, a 2.57 percent reduction in the mass of the lighter truck-based LTVs, slightly more in CUVs and minivans, slightly less in the heavier cars, almost double that in the heavier truck-based LTVs, and substantially less in the lighter cars – would be exactly safety-neutral as a point estimate: an estimated change of 0 ± 240 lives per year. In other words, just as in the 2010 report, any combination of mass reductions that maintain footprint and are proportionately somewhat higher for the heavier vehicles may well be safety-neutral or better as a point estimate and, in any case, may be very unlikely to significantly increase fatalities. The above estimates are offered only as computational examples, especially for the purpose of illustrating the width of the confidence bounds. The estimated safety effects that will appear in the FRIA, unlike these, will be based on the Volpe model, which forecasts the mass reductions of individual makes and models on a year-by-year basis.

The principal difference between the heavier vehicles, especially truck-based LTVs, and the lighter vehicles, especially passenger cars, is that mass reduction has a different effect in collisions with another car or LTV. When two vehicles of unequal mass collide, the delta V is higher in the lighter vehicle, in the same proportion as the mass ratio. As a result, the fatality risk is also higher. Removing some mass from the heavy vehicle reduces delta V in the lighter vehicle, where fatality risk is high, resulting in a large benefit, offset by a small penalty because delta V increases in the heavy vehicle, where fatality risk is low – adding up to a net societal benefit. Removing some mass from the lighter vehicle results in a large penalty offset by a small benefit – adding up to net harm. <u>These considerations drive the overall result</u>: fatality increase in the lighter cars, reduction in the heavier LTVs, and little effect in the intermediate groups. However, in some types of crashes that do not involve collisions between cars and LTVs, especially first-event rollovers and impacts with fixed objects, mass reduction is usually not harmful and often beneficial, because the lighter vehicles respond more quickly to braking and steering and are often more stable because their center of gravity is lower. Offsetting that benefit is the continuing historical tendency of lighter and smaller vehicles to be driven less well – although it continues to be unknown why that is so, and to what extent, if any, the lightness or smallness of the vehicle contributes to people driving it less safely.

Sensitivity tests

The preceding table of principal findings includes sampling-error confidence bounds for the five parameters used in the Volpe model: the statistical uncertainty that is a consequence of having less than a census of data. NHTSA's 2011 preliminary report acknowledged another source of uncertainty, namely that the baseline statistical model can be varied by choosing different control variables or redefining the vehicle classes or crash types, for example. Alternative models produce different point estimates. NHTSA believed it was premature to address that in the preliminary report. "The potential for variation will perhaps be better understood after the public and other agencies have had an opportunity to work with the new database."[13] NHTSA has now garnered 11 plausible alternative techniques that could be construed as sensitivity tests of the baseline model, which were tested or proposed by Farmer or Green in their peer reviews, Van Auken in his public comments, or Wenzel in his parallel research for DOE. The models use NHTSA's databases and regression-analysis approach, but differ from the baseline model in one or more terms or assumptions. NHTSA applied the 11 techniques to the latest databases to generate alternative Volpe-model coefficients. The range of estimates produced by the sensitivity tests gives an idea of the uncertainty inherent in the formulation of the models, subject to the caveat that these 11 tests are, of course, not an exhaustive list of conceivable alternatives. Here are the baseline and alternative results, ordered from the lowest to the highest estimated increase in societal risk per 100-pound reduction for cars weighing less than 3,106 pounds:

Fatality Increase (%) Per 100-Pound Mass Reduction While Holding Footprint* Constant

		Cars < 3,106	Cars ≥ 3,106	CUVs & Minivans	LTVs[†] < 4,594	LTVs[†] ≥ 4,594
Baseline estimate		**1.56**	**.51**	- .37	**.52**	- .34
95% confidence bounds (sampling error)	Lower:	.39	- .59	- 1.55	- .45	- .97
	Upper:	2.73	1.60	.81	1.48	.30
11 Alternative Models						
1. Track width/wheelbase w. stopped veh data		.25	- .89	- .13	- .09	- .97
2. With stopped-vehicle State data		.97	- .62	- .33	.35	- .80
3. By track width & wheelbase		.97	.24	- .24	- .07	- .58
4. W/O CY control variables		1.53	.43	.04	1.20	.30
5. CUVs/minivans weighted by 2010 sales		1.56	.51	.53	.52	- .35
6. W/O non-significant control vars		1.64	.68	- .46	.35	- .54
7. Incl. muscle/police/AWD cars/big vans		1.81	.49	- .37	.49	- .76
8. Control for vehicle manufacturer		1.91	.75	1.64	.68	- .13
9. Control for veh manufacturer/nameplate		2.07	1.82	1.31	.66	- .13
10. Limited to drivers with BAC=0		2.32	1.06	- .19	.86	- .58
11. Limited to good drivers[‡]		3.00	1.62	.00	1.09	- .30

*While holding track width and wheelbase constant in alternative model nos. 1 and 3.
[†]Excluding CUVs and minivans.
[‡]BAC=0, no drugs, valid license, at most 1 crash and 1 violation during the past 3 years.

For example, in cars weighing less than 3,106 pounds, the baseline estimate associates 100-pound mass reduction, while holding footprint constant, with a 1.56 percent increase in societal fatality risk. The corresponding estimates for the 11 sensitivity tests range from a 0.25 to a 3.00 percent increase. The sensitivity tests illustrate both the fragility and the robustness of the

---

[13] Kahane (2011), p. 81.

baseline estimates. On the one hand, the variation among the Volpe coefficients is quite large relative to the baseline estimate: in the preceding example of cars < 3,106 pounds, from almost zero to almost double baseline. That is so because the societal effect of mass reduction is small and, as Wenzel has said, it "is overwhelmed by other known vehicle, driver, and crash factors."[14] In other words, a variation in how to model some of those other vehicle, driver, and crash factors – which is exactly what the sensitivity tests do – can appreciably change the estimate of the societal effect of mass reduction.

On the other hand, the variations are not all that large in absolute terms. The ranges of the alternative estimates, at least these alternatives, are about as wide as the sampling-error confidence bounds for the baseline estimates. As a general rule, in the alternative models, as in the baseline models, mass reduction tends to be relatively more harmful in the lighter vehicles, more beneficial in the heavier vehicles. Thus, in all models, the point estimate of the Volpe coefficient is positive for cars < 3,106 pounds, and in all models except one, it is negative for LTVs ≥ 4,594 pounds. None of these models suggest mass reduction in small cars would be beneficial. All suggest mass reduction in heavy LTVs would be beneficial or, at worst, close to neutral. In general, any judicious combination of mass reductions that maintain footprint and are proportionately higher in the heavier vehicles is unlikely to have a societal effect large enough to be detected by statistical analyses of crash data.

Scope and limitations of the analyses

The many non-significant results in this report are not due to a paucity of data (except, perhaps, the paucity of very small or very light cars and LTVs during MY 2000-2007) or other weaknesses in the data, but because the societal effect of mass reduction while maintaining footprint, if any, is small. By contrast, a parallel statistical analysis of the effect of mass reduction with historically commensurate reductions of footprint (downsizing) shows strong, significant fatality increases in passenger cars.

Because the effects that need to be estimated are small, that raises some questions about the power of the results. First, can the estimates for the various crash types, most of which are individually not statistically significant, be combined to produce meaningful composite effects across crash types? The individual estimates are just intermediate computational tools used to obtain the composite effect; the key issue is the significance of the composite, not its component parts. Specifically, this report's analysis uses nine separate crash types and three vehicle types (as compared to 6 and 2 in the 2010 report) because it creates a better model, but the additional subdivision of the data further decreases the likelihood of significant results within the individual cells. A good analogy would be a health agency assigned to develop a general-purpose model that predicts the net risk increase due to any possible combination of different forms of tobacco consumption at different ages. Most of the individual effects – e.g., the risk increase due to smoking 100 cigars at age 39 – may be non-significant, but the composite effect of many cigarettes, cigars, and chewing tobacco over a 30-year period will likely be significant and accurately estimated by the model.

---

[14] Wenzel, T. (2011). *Assessment of NHTSA's Report "Relationships Between Fatality Risk, Mass, and Footprint in Model Year 2000-2007 Passenger Cars and LTVs – Draft Final Report."* (Docket No. NHTSA-2010-0152-0026). Berkeley, CA: Lawrence Berkeley National Laboratory, p. iv.

But unlike the tobacco model, even the composite effects are not significant in this report, except for cars < 3,106 pounds. What meaning, if any, do the non-significant estimates have? The regulatory analysis must provide the best estimate of the expected effect of mass reduction. The estimate has to be there, regardless of whether it is statistically significant – but with confidence bounds that indicate the range of uncertainty. (One reason that the regulatory analysis must have such an estimate is that it, too, is ultimately an intermediate computational tool in estimating the overall health and societal impact of CAFE and GHG regulation.) A good analogy would be a two-horse race. Analysis of past results and other data might not show a significant difference between the horses but it may still favor one horse over the other. A handicapper can definitely advise a client which horse is the better choice – just cannot give a 95 percent guarantee that the horse will win.

The estimates of this report are based on statistical analyses of historical data, which puts some limitations on their value for predicting the effects of future mass reductions. Analyses of historical data necessarily lag behind the latest developments in vehicles and in driving patterns because it takes years for sufficient crash data to accumulate. It is important to note that while the MY 2000-2007 database represents more modern vehicles with technologies more representative of vehicles on the road today than the previous report's MY 1991-1999 database, it still does not represent the newer vehicles that will be on the road in the 2017-2025 timeframe. The vehicles manufactured in the 2000-2007 timeframe were not subject to a footprint-based fuel-economy standard; vehicles actually became heavier on the average, not lighter during MY 2000-2007 and when they became heavier it was usually to provide additional features. NHTSA and EPA expect that the attribute-based standard will affect the design of vehicles such that manufacturers may reduce mass while maintaining footprint more than has occurred prior to 2017-2025. Therefore, it is possible that the analysis for 2000-2007 vehicles may not be fully representative of those vehicles that interact with the existing fleet in 2017 and beyond.

Statistical analyses can control for many factors such as a driver's age and gender, but there are other factors they do not control, such as driver characteristics that cannot be quantified with available demographic variables. The 11 alternative models described above shed some light on the sensitivity of the results to the way control variables are formulated, subject to the caveat that they are not an exhaustive list of alternatives. Furthermore, the analyses of this report are "cross-sectional": they compare the fatality rates for vehicles weighing n-100 pounds relative to other models weighing n pounds, rather than directly comparing the fatality rates for a specific make and model before and after a mass reduction had been implemented for the purpose of improving fuel economy. That type of direct statistical comparison, which might perhaps reveal different trends, will only become possible after substantial materials substitution has become more widespread in the vehicle fleet, not with the present data.

The estimates of the model are formulated for 100-pound mass reductions. What would be the effect of reducing mass by, say, 200 or 300 pounds? According to the model, if risk increases by 1 percent for 100 pounds, it would increase by 2 percent for 200 pounds and 3 percent for 300 pounds (more exactly, 2.01 percent and 3.03 percent, because the effects work like compound interest). Confidence bounds will grow wider by the same proportions.

For how many hundreds of pounds of mass reduction can the model predict accurately? This is the most difficult question. The model is best suited to predict the effect of a small change in mass, but everything else staying the same as it is now (MY 2000-2007 in CY 2002-2008). With each additional change from the current environment, the model may become somewhat less accurate. As stated above, the environment in 2017-2025 is bound to differ from 2000-2007. Nevertheless, one consideration provides some basis for confidence. This is NHTSA's fourth evaluation of the effects of mass reduction and/or downsizing, comprising databases ranging from MY 1985 to 2007. The results of the four studies are not identical, but they have been consistent up to a point. One of the most popular models of small 4-door sedans increased in curb weight from 1,939 pounds in MY 1985 to 2,766 pounds in MY 2007, a 43 percent increase. A high-sales mid-size sedan grew from 2,385 to 3,354 pounds (41%); a best-selling pickup truck from 3,390 to 4,742 pounds (40%) in the basic model with 2-door cab and rear-wheel drive; and a popular minivan from 2,940 to 3,862 pounds (31%). If the statistical analysis has, over the past years, been able to accommodate these gains on the order of 31-43 percent, perhaps it will also succeed in modeling the effects of mass reductions on the order of 10-20 percent, if they occur in the future.

In view of these considerations, what are the conclusions of the statistical analysis of MY 2000-2007 vehicles and the implications for the 2017-2025 fleet? NHTSA believes that only limited conclusions can be drawn from the statistical analysis. As stated above, the societal effect of mass reduction while maintaining footprint, if any, is usually small relative to the uncertainty in the statistics. The estimated effect of mass reduction in the 2000-2007 fleet is not statistically significant except in the cars < 3,106 pounds, where there is a significant increase. Estimates can be generated for the combined effects of mass reductions in various groups of vehicles, as required for the regulatory analysis of CAFE, with confidence bounds. In general, these estimates will not be statistically significant (except if mass reduction is limited to small cars, a strategy nobody appears to be considering). In other words, it cannot be concluded from the statistical analysis that mass reduction would have been harmful if it had been applied uniformly across the 2000-2007 fleet, let alone if it had been concentrated in the heavier vehicles. Downsizing (mass reduction with footprint reduction), by contrast, would have been harmful. Additional uncertainties are introduced if the results are used for predicting what might happen in the 2017-2025 fleet, since that future fleet will differ in various respects from the 2000-2007 fleet. During the final-rule deliberations, NHTSA expects to supplement the statistical results with fleet-simulation models that evaluate the crash safety of future light-weighted vehicles relative to baseline vehicle designs. Until then, the statistical results, by themselves, are not an unconditional warrant for mass reduction in the 2017-2025 fleet, but neither do they necessarily raise a red flag against mass reduction. Further research, combined with this analysis, will better help inform the agency's decision.

1. Background for a New Analysis of Fatality Risk, Mass, and Footprint

1.1 NHTSA's 2010 report

In April 2010, NHTSA issued a statistical analysis of relationships between fatality risk, mass, and footprint in MY 1991-1999 passenger cars and LTVs, based on CY 1995-2000 crash and vehicle-registration data. The text was incorporated in the safety chapter of the agency's final regulatory impact analysis (FRIA) of CAFE standards for MY 2012-MY 2016 passenger cars and light trucks. "Footprint" is defined as the wheelbase times the average of the front and rear track widths. It is a measure of a vehicle's "size."

NHTSA's previous 2003 weight-safety report had created a database of fatal crashes and vehicle registrations or VMT for model year 1991-1999 passenger cars and LTVs, permitting cross-sectional analyses of the fatality rate per million vehicle years or per billion miles by mass and/or size attributes, while controlling for driver age, gender, and other factors, in a single step of logistic regression.[15] For the 2010 report, NHTSA performed the logistic regressions with mass (curb weight) and footprint included as independent variables. Separate regressions were run for two vehicle classes – passenger cars (including 2-door and 4-door cars, but excluding police cars and high-performance "muscle" cars[16]) and LTVs (light trucks and vans, including in the 2010 report pickup trucks, SUVs, CUVs, minivan, and full-sized vans) – and six types of crashes: first-event rollovers, collisions with fixed objects, collisions with pedestrians/bicyclists/motorcycles, collisions with heavy vehicles, collisions with cars, and collisions with LTVs. Curb weight was formulated as a two-piece linear variable in order to estimate separate effects of mass reduction for cars < 2,950 pounds (the median mass of MY 1991-1999 cars in fatal crashes) and cars ≥ 2,950 pounds, likewise for LTVs < 3,870 pounds and ≥ 3,870 pounds. For a given vehicle class and weight range, the regression coefficients for mass in the six types of crashes were averaged, weighted by the number of baseline fatalities for the subgroup of MY 1996-1999 vehicles in CY 1996-2000. For example, in passenger cars < 2,950 pounds, mass reduction (while holding footprint constant) was beneficial in rollovers but harmful in the other types of crashes (although only slightly harmful in collisions with fixed objects or other cars). Footprint reduction (while holding mass constant) was harmful in all types of crashes except collisions with pedestrians. When the six individual mass coefficients were applied to baseline fatalities, the fatality increases added up to 301, which is a 2.21 percent increase over the baseline total of 13,208 societal fatalities in crashes involving cars < 2,950 pounds:

---

[15] Kahane, C. J. (2003). *Vehicle Weight, Fatality Risk and Crash Compatibility of Model Year 1991-99 Passenger Cars and Light Trucks*, NHTSA Technical Report. DOT HS 809 662. Washington, DC: National Highway Traffic Safety Administration, http://www-nrd.nhtsa.dot.gov/Pubs/809662.PDF, Chapter 3 (car) and Chapter 4 (LTVs).

[16] Police cars and muscle cars have exceptionally high fatality rates, compared to other cars of the same size and mass, because of unusual driving patterns. Given that police and muscle cars are relatively heavy and that, moreover, muscle cars tend to have small footprint (short wheelbase), the regression analyses might attribute the high fatality rates to mass or footprint rather than the unusual driving patterns; see Kahane (2003), pp. 41-42 and 171-173, Kahane (2010), pp. 483-486 and 512-514.

CAR WEIGHT GROUP	CRASH TYPE	BASELINE FATALITIES	EFFECT OF 100 LB RED (%)	FATALITY INCREASE	EFFECT OF .65 SQ FT FOOTPRINT RED (%)
CARS LT 2950 LBS	1ST EVENT ROLLOVER	995	- 1.59	-16	6.07
	FIXED OBJECT	3,357	.64	22	1.62
	PED/BIKE/MOTORCYCLE	1,741	3.23	56	- .13
	HEAVY TRUCK	1,148	3.98	46	1.14
	CAR	3,210	.99	32	.84
	LTV	4,091	3.95	162	1.04
	OVERALL	13,608	2.21	301	

Similarly, based on the regression analyses, Table 1-1 shows that the overall effects of reducing mass by 100 pounds while holding footprint constant were a 0.90 percent societal fatality increase in crashes involving cars ≥ 2,950 pounds, a 0.17 percent increase in LTVs < 3,870 pounds, and a 1.90 percent societal benefit in LTVs ≥ 3,870 pounds:

TABLE 1-1: 2010 REPORT[17] (MY 1991-1999 VEHICLES IN CY 1995-2000)
EFFECTS OF MASS REDUCTION WHILE MAINTAINING FOOTPRINT - ACTUAL REGRESSION RESULTS

	FATALITY INCREASE PER 100-POUND MASS REDUCTION (%)			
CRASH TYPE	CARS < 2,950	CARS ≥ 2,950	LTVs < 3,870	LTVs ≥ 3,870
1st-EVENT ROLLOVER	- 1.59	- 1.33	- 4.61	- 4.94
HIT FIXED OBJECT	.64	1.09	.08	- .55
HIT PEDESTRIAN/BIKE/MOTORCYCLE	3.23	- .60	.51	- .48
HIT HEAVY VEHICLE	3.98	.84	4.43	- .67
HIT CAR	.99	.74	- .17	- 1.78
HIT LTV	3.95	1.82	3.00	- 1.92
OVERALL	2.21±0.91	.90±0.91	.17±0.82	- 1.90±1.18

The 95% confidence bounds for the four overall estimates are each close to ±1 percentage point, as shown in Table 1-1. In other words, the fatality increase in the lighter cars and the societal benefit of mass reduction in the heavier LTVs are statistically significant; the estimates for the two intermediate vehicle classes are not. The FRIA used the four overall estimates in Table 1-1 to estimate the safety effects of likely mass reductions in MY 2012-2016 vehicles in response to the new CAFE standards, over the life of those vehicles.

NHTSA, however, had concerns that some of the actual regression results might be inaccurate or at least not indicative of the likely effect of future mass reductions. Specifically:

- Some of the coefficients for cars < 2,950 pounds may have been so strong because lighter cars in those years were poorly driven and had higher crash rates (see Section 1.7). That did not appear to be an effect of mass *per se* and might not persist with future mass reductions.

---

[17] Kahane (2010), last section of Table 2-15 on p. 514 and Table 3-2 on p. 525.

- Likewise, the strong effect for cars < 2,950 in pedestrian crashes might reflect the architecture of the cars rather than mass *per se* and might not persist in the future, depending on how vehicles are designed.
- The strong negative coefficients for LTV rollovers were volatile, possibly due to the near multicollinearity of mass and footprint and/or the inclusion of too wide a variety of vehicles in the analysis – as evidenced by strong <u>positive</u> coefficients when the same regressions were rerun for pickup trucks only.

Thus, NHTSA supplemented the overall estimates based purely on the actual regression results with an "upper-estimate" and a "lower-estimate" scenario that set aside some of the individual regression coefficients and replaced them with a range of estimates derived from other sources. The three alternative scenarios are shown in Table 1-2:

Table 1-2: Fatality Increase (%) per 100-Pound Mass Reduction While Maintaining Footprint

	Actual Regression Result Scenario	**Upper-Estimate Scenario**[18]	**Lower-Estimate Scenario**
Cars < 2,950 pounds	2.21	2.21	1.02
Cars ≥ 2,950 pounds	0.90	0.90	0.44
LTVs < 3,870 pounds	0.17	0.55	0.41
LTVs ≥ 3,870 pounds	-1.90	-0.62	-0.73

The FRIA estimated the safety effects of likely mass reductions in MY 2012-2016 by each of the three scenarios. All three supported the same conclusion: any reasonable combination of mass reductions in MY 2012-2016 vehicles – concentrated, at least to some extent, in the heavier LTVs and limited in the lighter cars – would not significantly increase fatalities and might well decrease them.

1.2 Next steps proposed in the 2010 final rule for 2012-2016 CAFE

The safety discussion in the May 7, 2010 preamble of the final rule for MY 2012-2016 light-duty vehicle GHG emission standards and CAFE standards ends by asking, "How do the agencies plan to address this issue going forward?"[19] The plan includes essentially three steps.

- "First, NHTSA is in the process of contracting with an independent institution to review the statistical methods that NHTSA and DRI have used to analyze historical data related to mass, size and safety, and to provide recommendation [*sic*] on whether the existing methods or other methods should be used for future statistical analysis of historical data."

The contract was awarded to the University of Michigan Transportation Research Institute (UMTRI). Paul E. Green was the principal investigator. The review comprised not only the NHTSA and DRI reports but approximately 20 other published studies of relationships between mass, size, and fatality risk. NHTSA also requested two additional peer reviews of just its 2010

---

[18] For passenger cars, the upper-estimate scenario <u>is</u> the actual-regression-result scenario.
[19] 75 Fed. Reg. 25324 (May 7, 2010); the discussion of planned statistical analyses is on pp. 25395-25396.

report from Charles Farmer of the Insurance Institute for Highway Safety (IIHS) and from Anders Lie of the Swedish Transport Administration. The three reviews, conducted by OMB guidelines, have been published in the NHTSA docket.[20] The comments and recommendations of the reviewers, discussed in Section 1.8 have been implemented in the analyses of both the preliminary and the current versions of this report.

- "Second, NHTSA and EPA, in consultation with DOE, intend to begin updating the MYs 1991-1999 database on which the safety analyses in the NPRM and final rule are based with newer vehicle data."

Soon after the publication of the 2010 report, NHTSA in collaboration with EPA and DOE began to develop new databases that start where the previous data ended. The previous database of MY 1991-1999 vehicles in CY 1995-2000 crashes has become outdated as new safety technologies, vehicle designs and materials were introduced and as the public's preferences in vehicles and their driving patterns shifted. The new databases comprise crash and exposure data of MY 2000-2007 cars and LTVs in CY 2002-2008. They are the most up-to-date possible, given the processing time for crash data and the need for enough crash cases to permit statistically meaningful analyses. NHTSA also worked in consultation with EPA and DOE to update the 2010 analysis methods in response to new vehicle technologies and the lessons on what worked and what did not work in 2010. The new databases and revised analysis methods are the basis for the preliminary version (November 2011) and for this final version of the report.

- Third, "The agencies intend to begin working collaboratively and to explore with DOE, [the California Air Resources Board] CARB, and perhaps other stakeholders ... to coordinate government supported studies and independent research, to the extent possible, to help ensure the work is complementary to previous and ongoing research and to guide further research in this area... Extending this effort to other agencies will help to ensure that all aspects of the weight-safety relationship are considered completely and carefully."

In May 2011, this effort by EPA, DOE and NHTSA has culminated in making NHTSA's making the databases used in the preliminary version of this report available to the public at, enabling researchers to analyze the same data and minimizing discrepancies in the results that would have been due to inconsistencies across databases. It also led NHTSA to host a workshop on the effects of mass and size on safety on February 25, 2011, bringing together experts in the field to discuss the questions that NHTSA must grapple with in the upcoming CAFE rulemaking.[21] The databases were subsequently updated in April 2012 in response to peer-review and public comments on the preliminary report and to add the most recent State crash data, as discussed in Section 2.5 and may be downloaded at http://www.nhtsa.gov/fuel-economy; researchers are encouraged to check http://www.nhtsa.gov/fuel-economy from time to time to ascertain they have the current version of the databases.

---

[20] Items 0003 (Lie), 0005 (Farmer) and 0022 (Green) in Docket No. NHTSA-2010-0152, which may be accessed by entering "NHTSA-2010-0152" at http://www.regulations.gov and clicking on "Search."
[21] http://www.nhtsa.gov/Laws+&+Regulations/CAFE-+Fuel+Economy/NHTSA+Workshop+on+Vehicle+Mass-Size-Safety.

1.3 NHTSA's November 2011 preliminary report

In November 2011, simultaneous with the NPRM for 2017-2025 CAFE and its PRIA, NHTSA issued its preliminary statistical analysis of relationships between fatality risk, mass, and footprint in MY 2000-2007 passenger cars and LTVs, based on CY 2002-2008 crash and vehicle-registration data that was available at that time.[22] The principal findings of the preliminary analysis are estimates of the overall percentage increases or decreases, per 100-pound mass reduction while holding footprint constant, in societal fatalities involving five classes of vehicles/weight ranges:

Table 1-3: November 2011 Preliminary Report (MY 2000-2007, CY 2002-2008)
Fatality Increase (%) per 100-Pound Mass Reduction While Holding Footprint Constant

	Point Estimate	95% Confidence Bounds
Cars < 3,106 pounds	1.44	+ .29 to +2.59
Cars ≥ 3,106 pounds	.47	- .58 to +1.52
CUVs and minivans	- .46	-1.75 to + .83
Truck-based LTVs < 4,594 pounds	.52	- .43 to +1.46
Truck-based LTVs ≥ 4,594 pounds	- .39	-1.06 to + .27

The preliminary analysis was published for public comment in Docket No. NHTSA-2010-0152. Charles Farmer, Paul E. Green, and Anders Lie, the researchers who reviewed the 2010 report, again peer-reviewed the November 2011 report.[23] The first three chapters of the current, final report by-and-large recapitulate the November 2011 analysis, differing only in the addition of a modest quantity of crash data that became available in the interim and a few revisions suggested in the peer-review and public comments. Essentially, every analysis in the November 2011 report from this point to the end of Chapter 3 is repeated with the updated database, with usually quite small changes in the results. The fourth chapter of this report addresses new issues raised in the peer and public reviews.

1.4 Developments in MY 2000-2007 vehicles

The most noticeable development during 2000-2007 was the great increase in crossover utility vehicles (CUV), which are SUVs of unibody construction, often but not always upon a platform shared with passenger cars, such as Ford Escape or Toyota RAV4. Appendix A lists the 2000-2007 vehicles that are considered CUVs in this report. Table 1-4 shows that CUVs increased from 1.59 percent of new light-vehicle sales in MY 2000 to 15.73 percent in 2007:

---

[22] Kahane, C. J. (2011, November). *Relationships Between Fatality Risk, Mass, and Footprint in Model Year 2000-2007 Passenger Cars and LTVs – Preliminary Report.* (Docket No. NHTSA-2010-0152-0023). Washington, DC: National Highway Traffic Safety Administration.

[23] Items 0035 (Lie), 0036 (Farmer) and 0037 (Green) in Docket No. NHTSA-2010-0152, which may be accessed by entering "NHTSA-2010-0152" at http://www.regulations.gov and clicking on "Search."

TABLE 1-4: PERCENT OF NEW LIGHT-VEHICLE SALES BY VEHICLE TYPE

	Cars			Truck-Based			Full-Size
	2-Dr	4-Dr	Pickup	SUV	CUV	Minivan	Van
2000	10.53	44.71	17.14	16.02	1.59	8.60	1.41
2001	10.39	42.11	18.14	15.17	6.08	6.66	1.44
2002	9.37	41.14	16.18	18.82	6.82	6.42	1.27
2003	8.20	41.18	16.85	17.52	8.95	5.93	1.36
2004	7.41	38.77	18.09	18.16	11.06	5.41	1.10
2005	6.82	41.01	15.89	13.88	13.41	7.71	1.28
2006	7.14	41.77	17.00	11.97	14.01	6.26	1.85
2007	7.33	43.97	14.01	13.17	15.73	4.60	1.19

But Table 1-4 also indicates that CUVs did not merely displace the more traditional truck-based SUVs, which in fact kept much of their market share. Some of the gain in CUVs may have come, directly or indirectly, from minivans and 2-door cars, both of which lost market share in 2000-2007. The statistics in Table 1-4 and throughout Section 1.4 are generated from the databases created for this report.

CUVs are somewhere between cars and trucks not only in their design and structure but also in their driving patterns: who and where. CUVs have the highest percentage of female drivers (63%), even more than minivans (61%), cars (57%), truck-based SUVs (53%), and pickup trucks (just 15%). CUVs also have the highest percentage of urban VMT, even more than cars and minivans. Urban female drivers have low fatality risk (see Section 3.2), which suggests that CUVs will also tend to have low fatality risk.

Heavier and larger, too: Another salient development in MY 2000-2007 is that the average vehicle became heavier: from 3,569 pounds in 2000 to a high of 3,916 pounds in 2004 and holding at 3,887 in 2007 (9% higher than in 2000). Average footprint also grew from 47.6 square feet in 2000 to 49.4 in 2007. Moreover, the various classes of vehicles became heavier and larger on the average, especially pickup trucks and truck-based SUVs:

- Cars: from 3,084 to 3,241 pounds (5%); from 44.0 to 45.2 square feet
- Pickup trucks: from 4,237 to 5,062 (19%); from 57.5 to 62.9 square feet
- Truck-based SUVs: from 4,244 to 4,925 (16%); from 46.2 to 51.4 square feet
- CUVs: from 3,362 to 3,847 (14%); from 42.3 to 46.8 square feet
- Minivans: from 3,852 to 4,306 (12%); from 51.7 to 53.7 square feet

A review of the specifications of individual makes and models during 2000-2007 shows that major redesigns usually added weight and size; "refreshing" between redesigns added new features; LTVs grew especially as consumers opted for stretched versions (such as crew cabs) and 4-wheel drive. The early CUVs tended to be relatively light and small, but by 2007 they came in all sizes. Growth within each vehicle type was only to a limited extent offset by the shift from truck-based SUVs and possibly minivans to somewhat lighter CUVs.

Safer: The 2000s were a decade of exceptional progress in vehicle safety. Unlike the vehicles in the databases in NHTSA's earlier studies, essentially all light vehicles in MY 2000-2007 were equipped with frontal air bags. Technologies that could meet Federal Motor Vehicle Safety Standards (FMVSS) or the voluntary agreement for LTV compatibility going into effect near the end of the decade already became widely available during 2000-2007:

	MY 2000	MY 2007
Electronic stability control (ESC)	5%	41%
Antilock brake system (ABS)	68%	86%
Any kind of curtain and/or side air bags	20%	66%
Curtain air bags	2%	62%
Curtains that deploy in rollovers	none	21%
Compatibility certification (truck-based LTVs)	45%	76%

But safety improvement was not limited to these specific technologies. Crash-test ratings issued as consumer information by NHTSA and the Insurance Institute for Highway Safety (IIHS) encouraged the design of vehicles to achieve good ratings. IIHS initiated offset-frontal testing in 1995. In MY 2000, still only 24 percent of new vehicles achieved a "good" overall rating, with 35 percent "acceptable," 20 percent "marginal," and 21 percent "poor." By MY 2007, 86 percent of new vehicles rated good, with 13 percent acceptable, under 1 percent marginal, and no poor ratings at all.[24]

The databases indicate that MY 2000-2007 CUVs had low societal fatality rates in CY 2002-2008. Without any adjustment for driver age and gender or driving environment, CUVs had a rate of just 10.3 societal fatalities per billion miles (i.e., counting not only their own occupants but also the occupants of the other vehicles in the crashes and any pedestrians or bicyclists) for MY 2000-2007 in CY 2002-2008, as compared to rates of 11.3 for minivans, 14.5 for cars, 16.5 for truck-based SUVs, and 21.8 for pickup trucks. Of course, without such adjustments (beyond the scope of this report), the rates are not directly comparable: pickup trucks, for example, are extensively driven in the most rural areas and primarily by males. Another indication of CUV safety is that 8.4 percent of their societal fatalities are in first-event rollover crashes, a proportion comparable to cars (6.8%) and minivans (7.1%). By contrast, 19 percent of the societal fatalities of truck-based SUVs were first-event rollovers.

But the truck-based SUVs themselves made great safety gains during MY 2000-2007. In MY 2000 SUVs, 21.5 percent of the societal fatalities were rollovers. By 2007, that proportion had dropped all the way to 4.3 percent, just as low as cars and CUVs. ESC, redesign with a lower center of gravity or wider track, curtains that deploy in rollovers, and increased belt use may all have contributed to that impressive achievement.

During MY 2000-2007 (and for some years before), many of the lightest and smallest vehicles were phased out, likely in response to dwindling sales and changing consumer preferences. Many poor safety performers, which were often holdovers from outdated designs and platforms,

---

[24] Bean, J.D., Kahane, C. J., Mynatt, M., Rudd, R.W., Rush, C.J., and Wiacek, C. (2009). *Fatalities in Frontal Crashes Despite Seat Belts and Air Bags*, NHTSA Technical Report. DOT HS 811 202. Washington, DC: National Highway Traffic Safety Administration, http://www-nrd.nhtsa.dot.gov/pubs/811102.pdf, pp. 6-8.

were phased out, likely in response to poor or marginal crash-test ratings. It so happens that those two groups extensively overlapped. Specifically, in model year 2000, new car models with sales totaling 1,543,000 had poor or marginal overall performance on the IIHS offset-frontal test: 1,276,000 of these cars weighed less than 3,106 pounds (the median mass of MY 2000-2007 cars involved in fatal crashes), whereas only 267,000 exceeded 3,106 pounds. By model year 2006, not a single new car had poor or marginal overall performance. In model year 2000, new truck-based LTV models with sales totaling 2,775,000 had poor or marginal overall performance on the IIHS offset-frontal test: 2,429,000 of these LTVs weighed less than 4,594 pounds (the 2000-2007 median in fatal crashes), whereas only 346,000 exceeded 4,594 pounds. By model year 2007, new LTVs with sales of just 79,774 had marginal performance and none had poor performance. In model year 2000, new CUV and minivan models with sales totaling 896,000 had poor or marginal overall performance on the IIHS offset-frontal test: 595,000 of these vehicles weighed less than 3,862 pounds (the 2000-2007 median), but only 301,000 exceeded 3,862 pounds. By model year 2006, not a single new CUV or minivan had poor or marginal overall performance.

1.5 Earlier studies

The key issue – mass versus "size" – has been variously perceived over the years. Soon after it became possible to statistically analyze large crash databases, researchers saw that lighter and smaller cars had higher fatality and injury rates – e.g., Mela's analysis of New York State data in 1974.[25] During the 1980s and 1990s, NHTSA and others pursued increasingly complex analyses that attempted to isolate the effect of car mass and size from other covariant factors such as driver age.[26] A shared feature of the early studies is that "mass" and "size" were to a large extent used interchangeably. There was less need to distinguish between mass and size because historic (especially 1975-1980) reductions in vehicle mass were accomplished by manufacturers reducing size when they redesigned a model, or by consumers simply retiring large, heavy cars and purchasing small, light cars of a different model. By 2002, the majority opinion of the National Academy of Sciences' expert panel was that "the downsizing and weight reduction that occurred in the late 1970s and early 1980s most likely produced between 1,300 and 2,600 crash fatalities and between 13,000 and 26,000 serious injuries in 1993."[27]

Nevertheless, researchers recognized mass and "size" as theoretically separate although historically confounded factors. Unlike mass, the right kind of "size" intuitively helps a vehicle without increasing harm to occupants of other vehicles in a crash. A wide track increases stability and reduces the likelihood of a rollover; crush space can protect a vehicle's occupants. A dissent by two panel members in an appendix to the NAS report, for example, argued that mass dissociated from size ought to have little influence on fatality risk except in determining the risk in one vehicle relative to another in a multi-vehicle collision – and even this has little net

---

[25] Mela, D. F. (1974). "How Safe Can We Be in Small Cars?" *International Congress on Automotive Safety, 3rd*, NHTSA Technical Report. DOT HS 801 481. Washington, DC: National Highway Traffic Safety Administration.
[26] NHTSA (1991). *Effect of Car Size on Fatality and Injury Risk*. Washington, DC: National Highway Traffic Safety Administration; Kahane, C. J. (1997). *Relationships Between Vehicle Size and Fatality Risk in Model Year 1985-93 Passenger Cars and Light Trucks*, NHTSA Technical Report. DOT HS 808 570. Washington, DC: National Highway Traffic Safety Administration, http://www-nrd.nhtsa.dot.gov/Pubs/808570.PDF.
[27] NAS (2002). *Effectiveness and Impact of Corporate Average Fuel Economy (CAFE) Standards*. Washington, DC: National Research Council, p. 77.

societal effect because, as one vehicle gets lighter and the risk for its own occupants increases, the risk will decrease for the occupants of the other vehicle by a more-or-less equal amount.

The issue became more directly relevant after 2000. The 2002 NAS report proposed restructuring CAFE standards in a way that would discourage harmful downsizing, for example, by setting higher CAFE targets for smaller vehicles rather than setting a universal standard applicable to the entire fleet of passenger cars or light trucks. In response, NHTSA developed footprint-based standards for MY 2008-2011 light trucks that were intended to discourage downsizing (by setting higher mpg levels for smaller footprints) but do not similarly discourage mass reductions that maintain footprint. Congress subsequently mandated an "attribute-based" approach for both passenger car and light truck CAFE standards in the Energy Independence and Security Act (EISA) of 2007.[28] Several technologies, most notably substitution of light, high-strength materials for conventional materials, have been proposed and in some cases implemented by vehicle manufacturers to reduce mass while maintaining not only footprint but also the structural strength of a vehicle.

The statistical analyses published by DRI in 2003 and 2005, often cited in the literature, have strongly supported the idea that mass and size can and should be analyzed independently. These regression analyses included curb weight, wheelbase, and track width as three separate independent variables and estimated an effect for each of them – unlike NHTSA's 1997 and 2003 analyses that use a single attribute, curb weight, which implicitly incorporated the size reductions that historically accompanied lower mass. DRI's analyses made it possible to estimate the effects of mass reduction <u>without</u> accompanying size reduction.[29] In fact, DRI's analyses estimated significant overall benefits for mass reduction in both passenger cars and LTVs if wheelbase and track width were maintained. Given the development of attribute-based standards and the prospect that materials substitution would allow future mass reduction without downsizing, NHTSA acknowledged it was essential to analyze mass and size independently and did so in its 2010 report. While the 2010 report did not show an overall benefit in all vehicles for mass reduction while maintaining footprint, it did show a benefit in some vehicles (heavy LTVs) as well as in some types of crashes – and it showed that mass reduction while maintaining footprint had a substantially more favorable impact than downsizing. In the meantime, NHTSA, DRI and others have performed many sensitivity tests on their databases and models, generating estimates that fall between the two initial sets of results.[30]

1.6 Hypothetical relationships between mass and societal fatality risk in the data

There is a strong historical trend of lighter vehicles having higher fatality rates for their <u>own</u> occupants. Two obvious factors contribute to the trend:

- Light vehicles have been, on the average, smaller than heavy vehicles and do not have the advantages of stability and crush space associated with large size.

---

[28] 49 U.S.C. § 32902(b).
[29] Van Auken and Zellner (2003); Van Auken and Zellner (2005a); Van Auken and Zellner (2005b).
[30] Van Auken and Zellner (2012a).

- In a collision between two vehicles, increasing the mass differential between the two vehicles (all else staying the same), increases the delta V for the lighter vehicle and thus also the risk for its occupants relative to the occupants of the heavier vehicle.

But the first factor might "drop out of the equation" if the analysis controls for size – e.g., by adding a size parameter such as footprint as an independent variable. The second factor might drop out if the dependent variable is the <u>societal</u> fatality rate including the fatalities in the partner vehicles – because the increase in fatality risk for the occupants of the light vehicle is offset by lower fatality risk for the occupants of the other vehicles in the collision. With these two factors out of the picture, would mass still have any residual <u>statistical</u> relationship with societal fatality risk in an analysis that controls for footprint – and in what direction?

Effects of conservation of momentum (delta V): In a collision of two light vehicles (cars or LTVs), reducing the mass of one of the vehicles would have increased its delta V and its occupants' risk, but it would have reduced both in the other vehicle. However, the two opposite effects do not necessarily cancel out to zero. When relatively light vehicles (e.g., cars < 3,000 pounds) collide with heavier ones (e.g., LTVs > 4,000 pounds), there are substantially more fatalities in the cars than in the LTVs. A further reduction in the mass of the cars would augment societal fatality risk, because an x percent increase in the many car-occupant fatalities would exceed in absolute terms the x percent reduction of the few occupant fatalities in the partner LTVs.[31] But mass reduction in the LTVs would diminish societal risk, because a y percent reduction in the many car-occupant fatalities would exceed in absolute terms the y percent increase of the few occupant fatalities in the LTVs. A safety-neutral effect can still be achieved by simultaneously reducing mass in both vehicles. But the statistical analyses of this report, which estimate effects of reducing mass in the case vehicle while the other vehicle remains unchanged, will tend to show, in collisions of two light vehicles, net harm for mass reduction in the lighter vehicles and net benefits for mass reduction in the heavier vehicles. This, more than any other factor will drive the current report's as well as the 2010 report's results by vehicle class – namely, overall harm for mass reduction in the lighter cars, overall benefit in the heavier LTVs.

There are, however, occasional situations where increased mass could benefit the occupants of the case vehicle without harming any other person. A heavy vehicle may be able to knock down a medium-size tree and continue moving forward, whereas a lighter vehicle would have come to a complete stop – and likewise for collisions with other partially moveable objects such as unoccupied parked vehicles, deformable poles, or large animals. This is not merely an academic point, as shown in Partyka's analysis of frontal impacts of passenger cars into trees or poles in NASS data: 56% of the heaviest cars significantly damaged the tree or pole, as compared to only 28-32% of the subcompact or compact cars.[32] "Significant damage to a tree or pole" includes cracking, shearing, or tilting a tree or pole; uprooting a tree; separating a pole from its base; or damage that resulted in replacement of the pole. In other words, extra mass reduced the car's delta V at least to some extent in approximately ¼ of the frontal collisions with fixed objects.

---

[31] Kahane (2003), pp. 105, 107, and 159; *New Crash Tests Demonstrate the Influence of Vehicle Size and Weight on Safety in Crashes*, IIHS News Release, April 14, 2009, http://www.iihs.org/news/rss/pr041409.html.
[32] Partyka, S.C. (1995). *Impacts with Yielding Fixed Objects by Vehicle Weight*. NHTSA Technical Report. DOT HS 808 574. Washington, DC: National Highway Traffic Safety Administration.

(Even here, there could be exceptions; it might be better for a guardrail to stop a vehicle completely than to let it go through, if there is something dangerous on the other side.)

Similarly, in a collision of a light vehicle with a medium-size truck (including LTVs with GVWR ≥ 10,000 pounds, not yet regulated by CAFE), additional mass in the light vehicle would make it transfer more of its momentum to the truck, reducing the light vehicle's ΔV and the fatality risk of its own occupants. (The fatality risk in the truck is so low that its slight increase in ΔV will not offset the benefit for the car's occupants.)

Energy absorption in single-vehicle crashes: In collisions with fixed objects, reducing mass of the vehicle while leaving its size and structural strength unchanged would translate to lower energy absorption required by the vehicle structure, which would tend to reduce fatality risk. Similarly, in rollovers, reducing a vehicle's mass while leaving its roof structure unchanged could reduce the force applied on the roof once the vehicle has overturned.

Benefits of enlarging footprint: Additional track width contributes directly to the static stability of a vehicle and its resistance to rollover. Wheelbase and track width are also protective because they enhance directional stability (preventing loss of control). Having more vehicle around the occupants could also create more potential crush space for the occupant's ride-down (an opportunity to enhance crashworthiness).[33]

Effects of mass on handling and stability: Adding mass to a vehicle while changing nothing else will make it slower to respond to steering, braking, or acceleration while reinforcing its tendency to proceed in a straight line at the same speed. Mass reduction through material substitution has the potential to enhance steering and braking capabilities, if the vehicle's brake and steering systems are left unchanged (or at least not reduced in capacity to an extent fully commensurate with the mass reduction). This enhanced performance and vehicle control would usually benefit crash avoidance. Specifically, when drivers initiate emergency maneuvers upon finding their vehicles out of control or pointed in the wrong direction, any extra mass could make it even more difficult for them to regain directional control.

Historically, mass has rarely been added or removed "while changing nothing else." Often, mass added by more comfortable interiors, luxury features, or more powerful engines has resulted in raising a vehicle's center of gravity (cg), making it more rollover-prone. But some features that add mass tend to lower the cg, such as 4-wheel drive equipment in some of the heavier LTVs. The statistical analyses of this report generally show a reduction of rollovers and impacts with fixed objects for lighter vehicles of the same footprint, consistent with the hypothesis that steering and braking capabilities are enhanced.

Factors historically correlated with mass: Several factors have been historically more correlated with mass than with footprint even though (unlike momentum conservation and braking/steering response) they are not features of mass *per se*. These factors will nevertheless contribute to the effects of mass reduction estimated in statistical analyses of historical data. Before the more widespread use of light, high-strength materials, less mass for the same footprint may have signified a structurally weaker vehicle. Potentially protective structure on the front

---

[33] Van Auken and Zellner (2005b), pp. 10-22.

and side of a vehicle beyond the wheels (overhang) adds mass without adding footprint. Similarly, raising a vehicle's sills for protection in side impact can add mass without footprint.[34] Historically, the frontal profile of small cars has been pedestrian-unfriendly: because the hood is short, the pedestrian's head is more likely to contact rigid structures such as the windshield header.[35] This is evidently not an issue of mass *per se*; nevertheless, higher rates of pedestrian fatalities continue to be statistically more associated with low mass than with small footprint.

Possible driver-vehicle interface factors: Historically (1976-2009), small, light vehicles have had higher collision-involvement rates (with or without injury) than larger, heavier vehicles of the same type, even after controlling for urbanization. In 1988, for example, the Highway Loss Data Institute (HLDI) reported that "small cars have consistently more injury and collision claims than large cars. This has been true for every year that HLDI has published insurance claim information [1976 onwards]."[36] A chart in HLDI's report showed that claims were more frequent for small cars than large cars within urban areas and likewise within rural areas. In 1998 HLDI announced, "Claims for crash damage are more frequent for small cars than for large ones"[37] and in 2009, "Small 4-door cars had higher frequencies than larger 4-door cars."[38] The higher incidence of smaller cars going out of control and running off the road explains some of this phenomenon. But in 1999-2000, 84 percent of cars' crash involvements (with or without injury) were collisions with other vehicles and less than 2 percent of those collisions involved loss of control[39] – yet small, light cars had higher crash rates there, too. It is unclear if the historical trends toward higher crash rates were primarily associated with size, mass or both; the studies did not attempt to isolate the effect of mass from size. For example, the HLDI reports compare various size classes of vehicles, but the vehicles larger size classes also tend to be heavier. The higher crash rates of small, light vehicles suggest there may be another factor – namely that, at least historically, for reasons that are not necessarily understood, small, light vehicles have not been driven as well as larger, heavier ones, as will be discussed in the next section.

1.7 Culpability in 2-vehicle crashes: relationships with mass and footprint

Key evidence to support the hypothesis that small, light vehicles are less well driven than larger and heavier vehicles of the same type comes from statistical analyses of who is culpable in 2-vehicle crashes. The data show that the lighter and smaller the vehicle, the more likely the driver of that vehicle was culpable. These are the findings of new analyses of the MY 2000-2007

---

[34] Kahane (2003), pp. 249-273 indicates the high fatality risk when light cars are hit in the side by LTVs and that the height mismatch (called D_AHOF in the report) accounts for a significant portion of the increased risk.

[35] Kahane (2003), pp. 98-99; Blodgett, R. J. (1983). *Pedestrian Injuries and the Downsizing of Cars*. Paper No. 830050. Warrendale, PA: Society of Automotive Engineers; MacLaughlin, T.F., and Kessler, J.W. (1990). *Pedestrian Head Impact Against the Central Hood of Motor Vehicles – Test Procedure and Results*. Paper No. 902315. Warrendale, PA: Society of Automotive Engineers.

[36] IIHS Advisory No. 5, July 1988, http://www.iihs.org/research/advisories/iihs_advisory_5.html.

[37] News Release, February 24, 1998, http://www.iihs.org/news/1998/hldi_news_022498.pdf.

[38] Auto Insurance Loss Facts, September 2009, http://www.iihs.org/research/hldi/fact_sheets/CollisionLoss_0910.pdf.

[39] NHTSA (2000). *Traffic Safety Facts 1999*. Report No. DOT HS 809 100. Washington, DC: National Highway Traffic Safety Administration, p. 71; Najm, W.G., Sen, B., Smith, J.D., and Campbell, B.N. (2003). *Analysis of Light Vehicle Crashes and Pre-Crash Scenarios Based on the 2000 General Estimates System*, Report No. DOT HS 809 573. Washington, DC: National Highway Traffic Safety Administration, p. 48.

FARS database created for this report as well as NHTSA's 2010 analyses of MY 1991-1999 databases from FARS and the General Estimates System (GES) of the National Automotive Sampling System (NASS). However, a review of these analyses suggests the results may be changing to some extent over time.

The new analyses use the FARS database of MY 2000-2007 case vehicles involved in fatal crashes during CY 2002-2008 (see Section 2.2) and logistic regressions with many of the variables described in Sections 3.1 and 3.2 (but here applied only to the FARS data, not the induced-exposure data as in Chapter 3). They examine the subset of crashes involving exactly two vehicles and no pedestrians or bicyclists. The case vehicle is a MY 2000-2007 car or LTV, but the "other" vehicle can be any type (including heavy trucks and motorcycles) and any model year.

"Culpability" is initially defined by the FARS variables for "driver contributing factors" (up to four coded for each driver). If the driver of the case vehicle has any of the codes indicating a specific action that may lead to a crash[40] (not merely a condition such as fatigue) and the other driver does not, the case vehicle is defined to be culpable. Conversely, if the driver of the other vehicle has any of these codes and the case driver has none, the other vehicle is culpable. If neither driver has any of the codes, a vehicle is also defined to be culpable if it is moving and hits a stationary vehicle or if it frontally impacts the rear of a vehicle that was not backing up. Examples of culpable vehicles include: the striking vehicle in a front-to-rear collision, being on the wrong side of the centerline prior to a head-on collision, encroaching on somebody else's lane, and failing to yield the right of way at an intersection or a left turn across traffic. The analysis is limited to cases where, according to the above criteria, one of the drivers is culpable and the other is not.

An empirical problem with fatal-crash data is the tendency of a surviving driver to blame the deceased driver – and this may influence the assignment of culpability when there is no physical evidence or witnesses to the contrary. For example, in the subset of 4,960 crashes where both vehicles were MY 2000-2007 cars or LTVs, one vehicle was judged culpable, and one driver died, the deceased driver was judged culpable 68 percent of the time. When a lighter and heavier vehicle collide, the driver of the heavier vehicle is more likely to survive – and blame the driver of the lighter vehicle, who did not survive. To avoid this potentially serious confounding, the analysis was further limited to 15,674 collisions of MY 2000-2007 "case" cars or LTVs with another vehicle in which:

- Both drivers died; or
- Neither driver died (i.e., only passengers died); or
- The "other" vehicle was a heavy truck (because the case-vehicle driver hardly ever survived, no matter how heavy the case vehicle was); or
- The other vehicle was a motorcycle (because the case-vehicle driver almost always survived, no matter how light the case vehicle was)

Two-vehicle collisions were subdivided into seven types:

---

[40] DR_CF codes 3, 6, 8, 26, 27, 28, 30, 31, 33, 35, 36, 38, 39, 44, 46, 47, 48, 50, 51, 57, 58, 79, or 87.

- Moving vehicle hit stationary vehicle
- Front-to-rear
- Head-on, both going straight, one vehicle in the wrong lane
- One vehicle changing lanes, encroaching on another vehicle going straight ahead
- Meet at intersection or right angle, neither vehicle turning, one fails to yield right of way
- One turning left, one coming straight the opposite way, one fails to yield right of way
- All others

Here are the results of the logistic regression of the last five collision types, combined, of the odds that the case vehicle was culpable, by curb weight of the case vehicle (LBS100, in hundreds of pounds), the case vehicle type (4-door car being the default), and many of the control variables discussed in Sections 2.2, 3.1, and 3.2, such as driver age and gender. Footprint is <u>not</u> a variable in this regression, which estimates the trend for lower mass with historically commensurate reductions in footprint (downsizing):

```
The LOGISTIC Procedure:
2-vehicle crashes excluding moving-to-stopped and front-to-rear, no 'blame-the-victim' issues

            Response Profile

 Ordered                    Total
   Value      CULPABLE      Frequency

       1             1           9074
       2             2           6600

          Model Fit Statistics

                                Intercept
                  Intercept          and
Criterion              Only    Covariates

AIC               21338.641     20329.699
SC                21346.301     20490.554
-2 Log L          21336.641     20287.699

        Testing Global Null Hypothesis: BETA=0

Test                 Chi-Square      DF      Pr > ChiSq

Likelihood Ratio      1048.9421      20         <.0001
Score                 1005.0274      20         <.0001
Wald                   942.0335      20         <.0001
```

Analysis of Maximum Likelihood Estimates

Parameter	DF	Estimate	Standard Error	Wald Chi-Square	Pr > ChiSq
Intercept	1	0.8319	0.1263	43.4043	<.0001
LBS100	1	-0.0218	0.00351	38.7197	<.0001
TWODOOR	1	0.1551	0.0738	4.4208	0.0355
CUV	1	-0.0656	0.0794	0.6829	0.4086
MINIVAN	1	-0.2824	0.0768	13.5181	0.0002
SUV	1	-0.1310	0.0676	3.7551	0.0526
REG_PKP	1	0.0117	0.0632	0.0345	0.8527
HD_PKP	1	-0.0813	0.1175	0.4787	0.4890
DRVMALE	1	-0.0675	0.0822	0.6741	0.4116
M14_30	1	0.0738	0.00754	95.9432	<.0001
M30_50	1	0.0120	0.00422	8.1139	0.0044
M50_70	1	0.0165	0.00495	11.1771	0.0008
M70_96	1	0.0901	0.00982	84.1299	<.0001
F14_30	1	0.0560	0.00880	40.4994	<.0001
F30_50	1	-0.00118	0.00511	0.0533	0.8174
F50_70	1	0.0281	0.00618	20.6535	<.0001
F70_96	1	0.0587	0.0134	19.2717	<.0001
ABS	1	-0.0405	0.0537	0.5682	0.4510
ESC	1	-0.1636	0.0752	4.7361	0.0295
AWD	1	-0.0203	0.0501	0.1639	0.6856
VEHAGE	1	0.0196	0.00813	5.8308	0.0157

A vehicle that is 100 pounds lighter than another vehicle, with historically commensurate smaller footprint is 2.18 percent more likely to be the culpable vehicle, after controlling for vehicle type, driver age and gender, and some other factors. This is a statistically significant trend to higher odds of culpability as evidenced by Wald chi-square of 38.72 for the LBS100 coefficient (3.84 or more indicates statistical significance at the 2-sided .05 level). The coefficients for the control variables indicate: 2-door cars are significantly more likely to be culpable than 4-door cars, minivans less; gender (DRVMALE) matters little; drivers of both genders are much more likely to be at fault for each year of age they are under 30 or over 70 (M14_30, M70_96, F14_30, and F70_96); ESC significantly reduces culpability; older vehicles are more often culpable (VEHAGE). Because half of the involvements in collisions with other vehicles are culpable involvements, a 2.18 percent increase in culpable involvements corresponds to a 1.09 percent increase in all involvements with other vehicles (assuming no change in the non-culpable involvements) per 100-pound downsizing.

Lighter and smaller vehicles are significantly more likely to be culpable in crashes: their drivers committed errors, such as failing to yield, that precipitated collisions. But when footprint is added to the regression variables (which already include curb weight) to allow separate estimates of the effects of "lighter" and "smaller," neither has a significant effect, in part because the standard errors of the coefficients increase as correlated variables are added. The coefficients for curb weight and footprint become:

Parameter	DF	Estimate	Standard Error	Wald Chi-Square	Pr > ChiSq
LBS100	1	-0.0105	0.00688	2.3212	0.1276
FOOTPRNT	1	-0.0137	0.00714	3.6767	0.0552

Basically, the new regression splits the effect for LBS100 in the first regression (reducing mass and footprint) about equally between mass (holding footprint constant) and footprint (holding mass constant). (But with neither effect statistically significant, it cannot be considered a precise division.) Furthermore, the standard error of the LBS100 coefficient increases from .00351 in the previous regression to .00688. Approximately the same split, with an additional increase in standard error occurs if the two variables track width and wheelbase are substituted for footprint:

Parameter	DF	Estimate	Standard Error	Wald Chi-Square	Pr > ChiSq
LBS100	1	-0.00927	0.00758	1.4936	0.2217
TRAKWDTH	1	-0.0150	0.0121	1.5218	0.2173
WB_MIN	1	-0.00547	0.00356	2.3686	0.1238

More detailed analyses of the MY 2000-2007 database show:

- About the same trend to higher culpability in lighter and smaller vehicles in cars, CUVs, and minivans (1.96% per 100 pounds) as in truck-based LTVs (2.28%); both are statistically significant.
    - In the analysis of cars, CUVs, and minivans, if mass is entered as a two-piece linear variables, the effect of mass reduction is approximately the same in the lighter and the heavier vehicles.
- Statistically significant trends to higher culpability in lighter and smaller vehicles in:
    - Head-on collisions, both going straight
    - Meeting at an intersection or at a right angle, neither vehicle turning
    - One turning left, one coming straight the opposite way
- The absence of such trends in moving-to-stationary, front-to-rear, and lane-changing collisions. Perhaps the lighter vehicle has an advantage of stopping or turning quicker and/or the smaller vehicle, being less conspicuous, is more likely to become a target.

The corresponding result in NHTSA's 2010 report was based on MY 1991-1999 passenger cars involved in two-vehicle fatal crashes in which both drivers died, neither driver died, or the other vehicle was a heavy truck or motorcycle. The database extended from CY 1991 through 2008, a file of 29,814 cases. The coefficients for curb weight (expressed as a two-piece linear variable) and footprint were:

```
2010 REPORT: FARS, CARS
                              Standard         Wald
   Parameter      DF   Estimate    Error    Chi-Square   Pr > ChiSq

   UNDRWT00        1    -0.0179    0.00868     4.2466       0.0393
   OVERWT00        1    -0.00960   0.00789     1.4826       0.2234
   FOOTPRNT        1    -0.00038   0.00866     0.0019       0.9651
```

The net trend of higher culpability with reduced mass and size is about the same as in the more recent vehicles. The difference is that the regression attributes the effect almost entirely to mass and hardly at all to footprint. Moreover, the mass effect is stronger in the lighter cars (< 2,950 pounds) than in the heavier cars. The log-odds of being the culpable party increases by an estimated 1.79 percent as cars < 2,950 pounds get 100 pounds lighter, after controlling for driver age and gender (2,950 was the median curb weight of cars in MY 1991-1999). The 2010 report's analysis of unweighted CY 1995-2000 GES (primarily nonfatal crashes) produced similar results, again attributing most of the effect to mass reduction in the lighter cars, not to footprint:

```
2010 REPORT: GES, CARS
                              Standard         Wald
   Parameter      DF   Estimate    Error    Chi-Square   Pr > ChiSq

   UNDRWT00        1    -0.0142    0.00600     5.6037       0.0179
   OVERWT00        1    -0.00323   0.00558     0.3335       0.5636
   FOOTPRNT        1     0.00910   0.00598     2.3146       0.1282
```

On the other hand, the 2010 report's analysis of LTVs (including CUVs and minivans) in FARS showed a balanced attribution of the effects to mass and footprint, similar to the MY 2000-2007 vehicles:

```
2010 REPORT: FARS, LTVs
                              Standard         Wald
   Parameter      DF   Estimate    Error    Chi-Square   Pr > ChiSq

   UNDRWT00        1    -0.0101    0.00742     1.8399       0.1750
   OVERWT00        1    -0.0168    0.00591     8.1165       0.0044
   FOOTPRNT        1    -0.00962   0.00410     5.4982       0.0190
```

The preceding analyses are all <u>statistical</u>; they indicate that lighter and smaller vehicles are less well driven, but they do not say why. One hypothesis ("self-selection") is that, for some reason, less effective drivers are more likely to choose lighter and smaller vehicles – but the lightness or smallness of the vehicles is not the cause of the ineffective driving. Another hypothesis ("driver-vehicle interface") is that certain aspects of lightness and/or smallness in a car or LTV give a driver a perception of greater maneuverability that ultimately results in driving with less of a "safety margin," for example weaving in traffic. That may appear paradoxical at first glance, as maneuverability is, in the abstract, a plus. But the situation is not unlike powerful engines that theoretically enable a driver to escape some hazards but in reality have long been associated with high crash and fatality rates.[41]

If lighter and smaller vehicles are driven less well, <u>regardless of the reason</u>, a cross-sectional statistical analysis of the historical data will associate a higher fatal-crash rate with lower mass and/or footprint, even after controlling for other factors. The effect is real in a statistical sense. But it is only important for predicting the effect of future mass or size reductions if the lightness or smallness somehow causes the ineffective driving (driver-vehicle interface). If the observed effect is primarily self-selection, if the entire fleet were to proportionally lose mass or footprint, it presumably would not make everyone's driving proportionally worse.

In summary, the 2010 report's trend toward higher odds of culpability in lighter and smaller vehicles persists in the new database. But its nature may be changing over time. In the earlier database, the effect in the lighter cars was primarily attributed to mass, not footprint. Now the effect appears to be more balanced between mass and footprint in all vehicles and less associated with mass reduction in particular. Perhaps recent safety improvements and better ratings of small, light vehicles are mitigating public perceptions that these vehicles are less designed for safety and tempering the possible self-fulfilling prophecy that drivers who care about safety do not pick such vehicles.

1.8   Peer-review recommendations on NHTSA's 2010 report

NHTSA's 2010 report was reviewed later that year by:

- Charles M. Farmer, Ph.D., Director of Statistical Services, Insurance Institute for Highway Safety, Arlington, VA[42]
- Paul E. Green, Ph.D., Assistant Research Scientist, Vehicle Safety Analytics, University of Michigan Transportation Research Institute, Ann Arbor, MI[43]
- Mr. Anders Lie, Specialist, Traffic Safety Division, Swedish Transport Administration, Borlange, Sweden[44]

The reviewers generally accepted the methodology of the 2010 report, which NHTSA retained to a large extent in its 2011 preliminary report. Here are some of the changes from the 2010 to the 2011 report that were influenced by the peer-review recommendations:[45]

- Make the complete databases available to the public.
- Add ESC as a control variable and, when computing baseline fatalities, take into account that the future fleet will be ESC-equipped.

---

[41] Robertson, L.S. (1991), "How to Save Fuel and Reduce Injuries in Automobiles," *The Journal of Trauma*, Vol. 31, pp. 107-109; Kahane, C.J. (1994). Correlation of NCAP Performance with Fatality Risk in Actual Head-On Collisions, NHTSA Technical Report No. DOT HS 808 061. Washington, DC: National Highway Traffic Safety Administration, http://www-nrd nhtsa.dot.gov/Pubs/808061.PDF, pp. 4-7.
[42] Docket No. NHTSA-2010-0152-0005.
[43] Docket No. NHTSA-2010-0152-0022.
[44] Docket No. NHTSA-2010-0152-0003.
[45] Kahane (2011), pp. 32-33 presents a more extensive discussion of the peer reviews and NHTSA's response.

- Distinguish between various subgroups of LTVs (e.g., CUVs) by making them separate vehicle classes with separate regression analyses.
- Further address the multicollinearity of weight and size variables.
- Estimate sampling errors in a way that takes the study design fully into account.
- Estimate the combined effect of reducing mass by 100 pounds (or by varying amounts) simultaneously in all classes of vehicles.
- Add crash-test ratings as a control variable.
- Consider drivers' alcohol use as a control variable or add analyses limited to sober drivers.

The same three researchers also reviewed NHTSA's 2011 preliminary report, simultaneous with a public comment period: Docket No. NHTSA-2010-0152, starting with item 0024; the peer reviews are items 0035 (Lie), 0036 (Farmer), and 0037 (Green). Chapter 4 of this report presents new analyses and discussion addressing the peer-review and public comments.

1.9 Limitations of statistical analyses of historical crash and exposure data

The statistical analyses – logistic regressions – of trends in MY 2000-2007 vehicles generate a set of estimates of the possible effects of reducing mass by 100 pounds while maintaining footprint. While these effects might conceivably carry over to future mass reductions, there are reasons that future safety impacts of mass reduction could differ from projections from historical data:

- The statistical analyses are "cross-sectional" analyses that estimate the increase in fatality rates for vehicles weighing n-100 pounds relative to vehicles weighing n pounds, across the spectrum of vehicles on the road, from the lightest to the heaviest. They do not directly compare the fatality rates for a specific make and model before and after a mass reduction had been implemented for the purpose of improving fuel economy (which was rare in MY 2000-2007). Instead, they use the differences across makes and models as a surrogate for the effects of actual reductions within a specific model; those cross-sectional differences could include trends that are statistically, but not causally related to mass.

- While statistical analyses can control for many factors such as a driver's age and gender, there are other factors they do not control. If, for example, riskier drivers tend to prefer lighter and smaller vehicles and if the characteristics of these drivers cannot be quantified with available demographic variables, the analysis would probably attribute the higher crash rates to the mass and size of the vehicles.

- Analyses of historical data lag behind the latest developments in vehicles because it takes years for sufficient crash data to accumulate for the newer vehicles. Vehicles became heavier on the average, not lighter, during MY 2000-2007 and if they became heavier it was usually to provide additional features. While there were examples of materials substitution to reduce mass of some components or structure without degrading a vehicle's performance, most makes and models did not yet exhibit year-to-year trends of decreasing overall mass.

- While the MY 2000-2007 database represents more modern vehicles with technologies more representative of vehicles on the road today than the previous report's MY 1991-1999 database, it still does not represent what vehicles will be on the road in the 2017-2025 timeframe. The vehicles manufactured in the 2000-2007 timeframe were not subject to a footprint-based fuel-economy standard. NHTSA and EPA expect that the attribute-based standard will affect the design of vehicles such that manufacturers may reduce mass while maintaining footprint more than has occurred prior to 2017-2025. Therefore, it is possible that the analysis for 2000-2007 vehicles may not be representative of those vehicles that interact with the existing fleet in 2017 and beyond.

2. New Databases to Study Fatalities Per Billion VMT

2.0 Summary

The objective of this study is to compare the fatality rates of MY 2000-2007 cars and LTVs during CY 2002-2008 on as "level a playing field" as possible, in order to discover the intrinsic difference in the safety of light and/or small-footprint versus heavy and/or large-footprint vehicles. The databases must include information about drivers' age and gender, and other factors that differ by vehicle weight or type, in order to allow adjustments for those differences. For example, since heavy cars have older drivers, on the average, than light cars, putting heavy and light cars "on a level playing field" requires computing fatality rates for heavy versus light cars for drivers of any specific age. Since pickup trucks are driven more in higher-risk rural areas than cars, a fair comparison of pickup trucks and cars requires computing both rural and urban fatality rates for each. Since some makes and models are driven more miles per year than others, it is appropriate to compare the fatality rates per mile rather than per registration year.

The databases of MY 1991-1999 vehicles in CY 1995-2000 crashes used in NHTSA's previous reports have become outdated. New safety technologies, vehicle designs and materials have been introduced. ESC has become much more widely available; high-strength steel and aluminum are in wider use. The public's preferences in vehicles and their driving patterns have shifted – e.g., an increase in CUVs. The new databases comprising MY 2000-2007 vehicles in CY 2002-2008 crashes are the most up-to-date possible, given the processing time for crash data and the need for enough crash cases to permit statistically meaningful analyses. NHTSA has also worked in consultation with EPA and DOE to improve the databases – by estimating vehicle miles of travel (VMT) more accurately, using crash data from additional States, defining control variables more consistently across States, and supplying information on the mass of the "other" vehicle in two-vehicle collisions.

The Fatality Analysis Reporting System (FARS) provides most of the information about fatal crashes needed for this study: the type of crash and number of fatalities, the age and gender of the driver(s), the time and location. No single database has comparable exposure information for the "denominators" needed to compute fatality rates. R.L. Polk's National Vehicle Population Profiles (NVPP) count the number of vehicles of a given make-model and model year registered in any calendar year. A file of odometer readings, also supplied by R.L. Polk, was used to derive estimates of annual VMT by make and model. State data on primarily nonfatal crashes, specifically, on "induced-exposure" crashes, allow classification of the mileage by age, gender, urban/rural and other characteristics corresponding to the FARS data. Induced-exposure crashes are involvements as the non-culpable vehicle, in a two-vehicle collision. The distribution of such involvements within a particular area is believed to be an essentially random sample of travel through that area. Accurate estimates of the curb weight and footprint of vehicles, as well as other attributes such as the presence of electronic stability control (ESC), antilock brake systems (ABS), and side or curtain air bags are assembled from several publications.

This chapter describes how the various sources are merged to generate a database of fatal crash involvements and a database of induced-exposure crash involvements for model year 2000-2007 vehicles in calendar years 2002-2008. The databases parse vehicle miles by vehicle mass, footprint, driver age, gender, urban/rural, ... and are suitable for direct use in logistic regressions to estimate fatality risk as a function of these variables. It also points out where the new databases, which are available to the public at http://www.nhtsa.gov/fuel-economy, differ from the files used in NHTSA's 2003 and 2010 reports. The databases have been updated since NHTSA's November 2011 preliminary report to add the most recent State crash data and update the estimates of annual VMT, as discussed in Sections 2.4 and 2.5. Even after publication of this final report, researchers are encouraged to check http://www.nhtsa.gov/fuel-economy from time to time to ascertain they have the current version of the databases.

2.1 Vehicle classification, curb weight, footprint, and other attributes

The Vehicle Identification Number (VIN) allows precise classification of vehicles and analysis of their body style and safety equipment. The VIN is known, with few missing data on FARS (fatal crashes) and 13 State files (induced-exposure crashes) available for analysis at NHTSA for all or most of calendar years 2002-2008: Alabama, Florida, Kansas, Kentucky, Maryland, Michigan, Missouri, Nebraska, New Jersey, Pennsylvania, Washington, Wisconsin, and Wyoming. The VIN itself, however, is not coded on Polk registration files, or listed in publications that specify curb weights.

NHTSA staff developed a series of VIN analysis programs in 1991 for use in evaluations of Federal Motor Vehicle Safety Standards and other vehicle safety analyses.[46] The programs are updated periodically. They were extended to model year 2007 in preparation for this study and are available to the public at http://www.nhtsa.gov/fuel-economy. Based entirely on the VIN, the programs identify a vehicle's make-model, model year and body type, and the type of restraint system for the driver and the right-front passenger. Each vehicle is assigned two five-digit codes: a fundamental vehicle group (that includes all of a manufacturer's vehicles of the same type and wheelbase, and runs for several years, until those vehicles are redesigned) and a specific make-model. For example, Chevrolet Cavalier and Pontiac Sunfire, for model years 1995-2005 are two make-models that comprise a single car group. For LTVs, NHTSA's VIN decoder generally assigns separate 5-digit make-model codes to the various cab/body styles and drive trains.[47] But for passenger cars, and for the few LTVs that FARS assigns "car-like" make-model codes (i.e., a zero in the hundreds place), NHTSA's VIN decoder uses codes similar to FARS; here, vehicles with conventional and all-wheel drive may have the same make-model code. Body styles of passenger cars, based on the VIN, are 2-door convertibles, 2-door coupe/sedans, 3-door hatchbacks, 4-door sedans, 5-door hatchbacks, and station wagons. LTV types are pickup trucks, crossover utility vehicles (CUV), truck-based SUVs, minivans, and full-sized vans. A CUV is an SUV of unibody construction, often but not always upon a platform shared with passenger cars. Appendix A lists the 2000-2007 vehicles that are considered CUVs in this report; other SUVs are considered truck-based SUVs in this report.

Whereas Polk NVPP data do not include the actual VIN, their VIN-derived classification variables suffice to define the fundamental vehicle group, specific make-model and body style/truck type as above,[48] and permitted the Polk NVPP data to be merged with FARS or State

---

[46] Kahane (1994), pp. 18-19.
[47] For example, a Ford F-150 4x2 pickup truck with 2-door cab is 12210; 4x4 with 2-door cab, 12211; 4x2 extended cab, 12212; 4x4 extended cab, 12213; 4x2 crew cab, 12214; and 4x4 crew cab, 12215.
[48] But not the reverse: Polk's more detailed classification variables cannot be derived from these NHTSA codes.

crash data. NVPP data specify the number of vehicles registered as of July 1 of every calendar year. The file of odometer readings supplied by Polk classifies vehicles by the same variables as NVPP.

"Curb weight" is the weight of a ready-to-drive vehicle with a full tank of fuel and all other fluids, but no driver, passengers or cargo (as opposed to the "shipping weight," that excludes some fluids, and the "gross vehicle weight rating," that includes the vehicle and its permissible maximum load of occupants and cargo). Curb-weight information is derived from five sources:

1. 2000-2007 *Branham Automobile Reference Books*, Branham Publishing Co., Santa Monica, CA (nearly all cars and LTVs)
2. R.L. Polk's NVPP database (cars only)
3. FARS cases (most cars and minivans, many CUVs and SUVs, some full-sized vans, but no pickup trucks; FARS now usually specifies curb weights, and not shipping weights as it did in some earlier years)
4. www.cars.com
5. www.motortrend.com

All of these, in turn derive from the same original source: manufacturers' official weights for vehicles of a specified make-model and subseries (and, perhaps, engine + transmission), with all equipment standard for that subseries, but without any additional, purely optional equipment.

Branham's weights are usually quite detailed and complete. They were the primary source – the basis for 77 percent of the curb-weight estimates. Corresponding to any one of NHTSA's specific 5-digit vehicle-group and make-model codes and body style, Branham may list a single curb weight or a range. If Branham specifies a range of weights, however, it is not obvious how to identify the average within that range. Therefore, if NVPP provides the same or nearly the same range as Branham, but also provides a count of registrations for each figure in the range, the registration-weighted average curb weight in the NVPP is used. Likewise, if FARS (but not NVPP) provides a range similar to Branham, the N-of-crashes-weighted average is employed. NVPP and FARS data each accounted for just fewer than 10 percent of the curb-weight estimates. Among the 4 percent of vehicles where Branham specifies weights that seem inconsistent with the preceding or following model year or out of line with similar vehicles (e.g., the same LTV but with a different drive system or cab style), www.cars.com or www.motortrend.com may be consulted, or, for example, the average of the preceding and following model year's Branham weight may be substituted for that year's weight.[49] When Branham specifies a range of weights, and neither NVPP nor FARS offer a reliable average, a

---

[49] Or if, for example, year after year a 4x4 model weighs 400 pounds more than the 4x2 model, and in one model year they are listed as having the same weight, 400 pounds is added to the weight for the 4x4 model if the weight for the 4x2 model appears consistent with the previous and following years' 4x2 models.

central point in the Branham range is identified based on the statistics for those makes and models where Branham has a range and NVPP or FARS do have a reliable average.[50]

Published, manufacturer-defined weights for vehicles usually include only standard equipment, but NHTSA's compliance- and crash-test contractors weigh "typical" vehicles (including popular options) from the stock of retail dealerships. However, NHTSA's 2003 report showed that already by the 1990s, the average discrepancy between measured and published curb weights had shrunk to an average of 1 percent in cars and 2 percent in LTVs. That is because once-optional features such as automatic transmissions and air-conditioning have increasingly become standard equipment on entire make-models or subseries of them.[51]

Footprint is a measure of a vehicle's size, defined as the wheelbase times the average of the front and rear track widths. The Department of Energy (Wenzel) gathered measurements of wheelbase and track width from motortrend.com into a spreadsheet and provided it to NHTSA. Unlike NHTSA's 2010 report, footprint is not assumed to be identical for all vehicles of the same 5-digit car or LTV group. Track widths may vary slightly for vehicles built on the same or similar platforms. In the databases, wheelbase and track widths are measured in inches, footprint in square feet.

The other vehicle attributes included in the databases are:

- ABS (4-wheel)
- ESC
- Side air bags, including:
    - Curtain air bags
        - Rollover curtain bags that deploy and stay inflated in rollover crashes
    - Torso bags
    - Combination bags that provide torso and head protection
- Voluntary vehicle-to-vehicle compatibility certification for pickup trucks and SUVs[52], including:
    - Option 1: the primary energy-absorbing structure is low enough to adequately overlap the bumper height of passenger cars without additional structure
    - Option 2: a secondary energy-absorbing structure, often called a "blocker beam," below the primary structure
- All-wheel or 4-wheel drive (AWD)

---

[50] In general, these statistics show that the wider the range of weights (bran_lo to bran_hi) specified in Branham, the relatively closer the registration- or N-of-crashes-weighted average, branwt is to bran_lo – because, typically, sales are somewhat lower for the premium subseries or high-performance engines than for the more basic subseries. Based on GLM analyses of car models where Branham and NVPP or FARS weights are available, if bran_hi-bran_lo is exactly 124 pounds, branwt is .424 of the way up the range (53 pounds up from bran_lo) if the vehicle is an LTV and .325 of the way up (40 pounds up from bran_lo) if it is a car. If the bran_hi-bran_lo is greater or less than 124 pounds, branwt is proportionately (.039*log(nurange/124)) less or more of the way up, where nurange = bran_hi – bran_lo, but not less than 20 or more than 398 pounds. For example, for a car with bran_hi-bran_lo=200, branwt-bran_lo = 200x(.325-.039log(200/124)) = 61.27; for a car with bran_hi-bran_lo=500 (if such a car existed), branwt-bran_lo = 500x(.325-.039log(398/124)) = 139.76

[51] Kahane (2003), pp. 18-19.

[52] "Enhancing Vehicle-to-Vehicle Crash Compatibility, Commitment for Continued Progress by Leading Automakers," Docket No. NHTSA-2003-14623-0013.

- Insurance Institute for Highway Safety (IIHS) test results for frontal-offset and side impact

The ABS, ESC, AWD, and side-air-bag variables are coded 1 if the feature is standard equipment for that VIN, 0 if it is not available, or a number between 0 and 1 if it is optional but not decodable from the VIN. That number is the percent of vehicles, for that make, mode, and model year equipped with the feature. Information about these features was gleaned from www.safercars.gov and National Insurance Crime Bureau vehicle-identification manuals (models and subseries where standard or available) and Ward's Automotive Yearbooks (percent equipped if optional and not VIN-decodable). Note that vehicles may be coded 1 for up to three of the four side-air-bag variables, namely if they have curtain plus torso bags and the curtains deploy in rollovers. The variable on vehicle-to-vehicle compatibility was simply coded 1 (for Option 1) or 2 (for Option 2), as these features are either standard or unavailable. AWD is coded 1 for either all-wheel drive or 4-wheel drive.

AWD and 4x4 usually add substantial mass to a vehicle, relative to the same model with 2-wheel drive. However, the vast majority of 4x4 and AWD vehicles are LTVs where NHTSA's VIN decoder assigns separate make-model codes to 4x2, 4x4, and AWD – and each of these make-models has a different curb weight. For the relatively few passenger cars and LTVs (mostly CUVs) with "car-like" make-model codes that are available with either FWD/RWD or AWD, there is initially only one "central" curb weight defined for that make-model, NHTSA analyzes the VIN; if it is an AWD vehicle, AWD=1 and the curb weight is augmented; if not, AWD=0 and the curb weight is diminished.[53] IIHS ratings are available to the public at www.iihs.org; IIHS also supplied a spreadsheet of the ratings to NHTSA. Whereas the other variables are defined for all MY 2000-2007 vehicles, some makes and models have not been tested and have no ratings.

NHTSA's databases, which are available to the public at http://www.nhtsa.gov/fuel-economy, show the curb weight, footprint, and other attributes of each crash-involved vehicle. Appendices B and C of this report are codebooks for the variables on the public databases.

2.2 Fatal crash involvements: FARS data reduction

The preparation of the database of vehicles involved in fatal crashes consists of identifying: (1) the vehicle's make-model, body style, and curb weight, based on VIN analysis as described in the preceding section; (2) the type of crash, depending on the types and curb weights of other vehicles involved and whether non-occupants were involved; (3) the dependent variable, the count of fatalities in the crash (including fatalities in other vehicles and non-occupants); (4)

---

[53] A review of Branham weights for 8 make-models of cars available with FWD/RWD or AWD showed that AWD added an average of 168 pounds; that the AWD models averaged 115 pounds heavier than the "central" curb weight for that make-model and the FWD/RWD models, 53 pounds lighter. Thus, 115 pounds were added to the initial, central curb weight for cars with AWD=1 (if AWD was optional, not standard for that make-model) and 53 pounds were subtracted from the initial curb weight if AWD=0 (for those make-models where AWD was optionally available). Similarly, a review of Branham weights for 8 make-models of CUVs with car-like make-model codes and optional AWD showed that AWD added an average of 195 pounds; the AWD-equipped vehicles averaged 116 pounds more than the "central" curb weight; and the non-AWD vehicles averaged 79 pounds less than the "central" curb weight.

potential control variables, factors that correlate with both vehicle weight and fatality risk, such as driver age, urban/rural, etc.

The 2002-2008 FARS files contain 113,248 records of crash-involved vehicles of model years 2000-2007 with decodable VINs that can be assigned a curb weight and identified as passenger cars or LTVs (pickup trucks, CUVs, truck-based SUVs and vans, excluding incomplete vehicles[54] but including "300-series" pickups and vans with GVWR sometimes over 10,000 pounds). The database of fatal crash involvements consists of those 113,248 records. Table 2-1 assigns the 113,248 "case" vehicle records to nine basic crash types:

TABLE 2-1: FATAL-CRASH INVOLVEMENTS OF MY 2000-2007 CARS AND LTVs WITH KNOWN CURB WEIGHTS ON CY 2002-2008 FARS

	Passenger Cars	Truck-Based LTVs	CUVs & Minivans
1. First-event rollovers	2,976	6,395	660
2. Hit fixed object	10,575	7,437	1,319
3. Hit pedestrians/bikes/motorcycles	7,127	6,931	1,904
4. Hit heavy truck or bus	3,556	2,528	680
5. Hit car, CUV or minivan < 3,082 lbs.	4,701	5,939	1,245
6. Hit car, CUV or minivan ≥ 3,082 lbs.	5,133	5,167	1,211
7. Hit truck-based LTV < 4,150 lbs.	3,390	2,810	622
8. Hit truck-based LTV ≥ 4,150 lbs.	3,876	2,290	685
9. All other crash involvements	11,336	9,945	2,810
	52,670	49,442	11,136

The 24,091 "all other" crash involvements in Table 2-1 include:

- 19,802 in collisions that involved three or more vehicles
- 1,160 in single-vehicle crashes where it is difficult to tell if the first truly harmful event was a rollover or a collision with a fixed object
- 1,163 crashes involving fatalities to both occupants and non-occupants, or where it was not clear which of the two involved vehicles hit the fatally injured non-occupant
- 742 non-collisions of other types, such as first-event immersion or falling from a moving vehicle
- 652 two-vehicle collisions where the other vehicle was of other/unknown type or unknown mass
- 327 collisions with trains
- 245 collisions with animals, working vehicles, or on-road objects

---

[54] Although Branham may list curb weights for incomplete vehicles such as cab-chassis or RV cutoffs, it is unknown how much additional weight (and it may be a lot) is added during the second stage of building the vehicle – e.g., adding the RV body.

"First-event rollovers" include single-vehicle crashes where the rollover seemed to be the first truly harmful event, even if FARS coded an apparent tripping mechanism, such as a ditch, as the "first" harmful event. "Fixed-object" collisions include single-vehicle crashes where the case vehicle first left the travel lanes and then struck a substantial fixed object (including a parked car), regardless of whether it subsequently rolled over or not.[55] The third crash type includes collisions of one car or LTV with pedestrian(s), bicyclist(s), or motorcyclist(s), where the fatalities are not in the car or LTV, plus crashes involving two vehicles and non-occupant(s) where FARS clearly specifies that the case vehicle first hit and fatally injured the non-occupant, then hit the other vehicle without any additional fatality. Crash type 4 includes collisions involving one car or LTV and one or more heavy vehicles. Types 5-8 are limited to 2-vehicle collisions where the case vehicle is a 2000-2007 car or LTV and the "other" vehicle is a car or LTV, respectively, of known curb weight but any model year, not necessarily 2000-2007. Note that truck-based LTVs have relatively high proportions of first-event rollovers and collisions with light cars, CUVs or minivans (where most of the fatalities are in the car, CUV, or minivan).

The classification of vehicles and crash types differs from NHTSA's 2003 and 2010 reports primarily as follows:

- CUVs (which only began to appear in the late 1990s) as well as minivans differ from truck-based LTVs but are not exactly like cars, either. They will be classified here as a third, separate group of case vehicles, rather than included with other LTVs.[56]
- The groups of collisions with other light vehicles are limited to 2-vehicle collisions where the curb weights of both vehicles are known. Moreover, the old categories, "hit car" and "hit LTV" are each subdivided according to the weight of the other vehicle, above and below 3,082 or 4,150 pounds, respectively (the median weights of "other" vehicles in the database). In addition, CUV and minivan "other" vehicles have been grouped with cars, because they are relatively unaggressive in crashes, like cars and unlike truck-based LTVs.[57] Curb weights are now known for almost all "other" light vehicles on FARS because few are, at this point, so old that documentation is unavailable (e.g., from earlier weight-safety studies). The advantage of subdividing the "other" vehicles by curb weight is that, intuitively, the effect of mass reduction in the case vehicle would likely be different if the other vehicle is of approximately the same mass or of substantially greater mass.
- The "all other crashes" group is retained rather than excluded from further study.

The following driver, road, and environmental factors are control variables for "case" vehicles defined directly from FARS data:

DRVAGE – Driver age (range 14 to 96): include if AGE of the driver of the case vehicle is 14 to 96. Delete case if AGE=97 (97 or older), 99 (unknown), less than 14, or if no driver record exists.

---

[55] Rollovers preceded by collisions with devices such as guardrails, which might be solid impacts in some cases and mere tripping mechanisms in others, are classified in the "all other crashes" group.
[56] Full-size vans as well as the Chevrolet Astro/GMC Safari (although they are called minivans) remain in the truck-based LTV group.
[57] Kahane (2003), Chapter 6.

DRVMALE – Driver male (values 0, 1): if, for the driver of the case vehicle, SEX=1 (male) then DRVMALE=1, else if SEX=2 (female) then DRVMALE=0, else delete the case

NITE – Crash happened between 7:00 P.M. and 5:59 A.M. (values 0, 1): if HOUR = 6-18 (i.e., 6:00 a.m. – 6:59 p.m.) then NITE = 0, else if HOUR = 0-5 or 19-24 then NITE = 1, else delete the case

RURAL – Crash happened in a county with population density < 250 per square mile (values 0, 1): the Department of Energy (Wenzel) proposed this approach and supplied a list of population densities for every county in the United States, compatible with FARS (FIPS) State/county codes. If population density < 250 then RURAL=1, if ≥ 250 then RURAL=0, if COUNTY is unknown, delete the case. With the boundary at 250, 57 percent of the fatalities in 2002-2008 FARS are "rural," almost identical to the 58 percent rural based on the FARS variable ROAD_FNC. The advantages of county-based RURAL is that the definition is consistent across States and also between FARS and State crash data – whereas the interpretation of ROAD_FNC (the variable used in the 2003 and 2010 reports) may vary across States and may also differ from the way rural/urban is defined in some State files (and is not defined at all in other State files).

SPDLIM55 – Crash happened on a road with speed limit 55 or more (values 0, 1): if the accident-level variable SP_LIMIT if 55, 60, 65, 70, 75, or 80 then SPDLIM55 = 1; otherwise SPDLIM55 = 0 (includes speed limits 5-50, unknown, or stray values).

CY – Calendar year of the crash, range 2002 to 2008

VEHAGE – Age of the case vehicle, CY-MY, range 0 (for a new vehicle) to 8 (MY 2000 in CY 2008). Delete cases with CY-MY = -1.[58]

HIFAT_ST – Crash happened in a State with a higher-than-average fatality rate (values 0, 1): if the State had a higher-than-national-average overall fatality rate per million vehicle years, HIFAT_ST = 1, else 0. The fatality rate is the sum of 2002-2008 traffic fatalities, divided by 2008 registered vehicles, as tabulated by the Federal Highway Administration.[59] The 26 States with lower-than average rates are Alaska, California, Connecticut, Delaware, Hawaii, Illinois, Indiana, Iowa, Maryland, Massachusetts, Michigan, Minnesota, Nebraska, New Hampshire, New Jersey, New York, North Dakota, Ohio, Oregon, Pennsylvania, Rhode Island, Utah, Vermont, Virginia, Washington, and Wisconsin. The 25 jurisdictions with higher-than-average rates are Alabama, Arizona, Arkansas, Colorado, D.C., Florida, Georgia, Idaho, Kansas, Kentucky, Louisiana, Maine, Mississippi, Missouri, Montana, Nevada, New Mexico, North Carolina, Oklahoma, South Carolina, South Dakota, Tennessee, Texas, West Virginia and Wyoming. HIFAT_ST is essentially a geographical variable. The HIFAT_ST = 1 group is primarily a contiguous area consisting of the South, the Mountain States and the adjacent States Kansas and Missouri, characterized by substantial non-metropolitan populations and/or short winters. The HIFAT_ST = 0 group is primarily the Northeast, the Midwest (except Kansas and Missouri), and the Pacific States, characterized high urbanization, by long winters, and/or aging populations.

---

[58] Because corresponding exposure data might not be available. For example, if the new model year started selling October 1, there would be zero registrations in the NVPP file as of July 1.
[59] Table MV-1, State Motor-Vehicle Registrations – 2008, http://www.fhwa.dot.gov/policyinformation/statistics/2008/xls/mv1.xls.

2.3 Vehicle registration years: Polk NVPP data reduction

R.L. Polk's *National Vehicle Population Profile* (NVPP) databases do not include the actual VIN, but their VIN-derived variables suffice to define the fundamental vehicle group, specific make-model and body style/truck type as described in Section 2.1.[60] NVPP data specify the number of vehicles registered as of July 1 of every calendar year, and provide estimates of vehicle registration years by MY, CY, vehicle group, make-model, body style/truck type and, where needed, by State. NVPP data have no information, for example, on the age or gender of the drivers, or the annual VMT, or whether the vehicles were driven by day or at night.

2.4 Annual VMT: odometer readings from Polk

Fatality rates per hundred million vehicle miles of travel (VMT), rather than per million registration years, are the most widely accepted measure of risk. Estimates of the average VMT for a vehicle of a specific make-model and MY during a specific CY were derived as follows.

NHTSA estimated (for use in its regulatory analyses of lifetime costs and benefits) the number of miles that the average car or LTV was driven per year circa 2001, from the day the first owner acquires the vehicle, based on data from the Federal Highway Administration's 2001 National Household Transportation Survey (NHTS).[61] Table 2-2 shows the estimates for the first 10 years:

TABLE 2-2: AVERAGE ANNUAL VMT BY NUMBER OF YEARS IN SERVICE

	Cars	LTV	Average	Cumulative
1st year	14,231	16,085	15,158	15,158
2nd	13,961	15,782	14,872	30,030
3rd	13,669	15,442	14,556	44,586
4th	13,357	15,069	14,213	58,799
5th	13,028	14,667	13,848	72,647
6th	12,683	14,239	13,461	86,108
7th	12,325	13,790	13,058	99,166
8th	11,956	13,323	12,640	111,806
9th	11,578	12,844	12,211	124,017
10th	11,193	12,356	11,775	135,792

---

[60] But not the reverse: Polk's more detailed classification variables cannot be derived from NHTSA's fundamental vehicle group, specific make-model and body style/truck type.

[61] Lu, S. (2006). *Vehicle Survivability and Travel Mileage Schedules*, NHTSA Technical Report. DOT HS 809 952. Washington, DC: National Highway Traffic Safety Administration, http://www-nrd nhtsa.dot.gov/Pubs/809952.PDF; the numbers in Table 2-2 are the "updated [to 2001] results" from Tables 7 and 8 in Lu's report. Table 2-2 has changed from NHTSA's preliminary report, Kahane (2011), p. 43, where it was based on the "current" results from Tables 1 and 2 in Lu's report. However, those results were no longer current in the 2000s, but derived from a 1991 survey by the Energy Information Administration, showing lower VMT in absolute terms and also a faster drop-off in VMT as vehicles aged.

Table 2-2 does not correspond exactly to the average mileage for vehicles of a specific MY in a particular CY, because owners may acquire their new vehicles on any day of the year. For example, one customer bought his MY 2007 car on May 1, 2007 and only drove it for 7 months during CY 2007, while another bought her MY 2007 car back on November 1, 2006 and was already into her second year of ownership after November 1, 2007.

The calculations in Table 2-3 track the cumulative mileage of a simulated 10,000-vehicle run of a generic make-model in model year N. They assume the vehicles were sold at a uniform rate from October 1 of the previous calendar year (N-1) through September 30 of calendar year N; that each individual owner, starting from the date he or she buys the vehicles, drives at a constant rate of 15,158 miles per year until one year after the purchase date, then at a constant rate of 14,872 miles per year until two years after the purchase date, and so on; and (for simplicity) that no vehicles are retired.

TABLE 2-3: COMPUTATION OF VMT FOR A SPECIFIC MY IN A SPECIFIC CY DIVIDED BY THE NUMBER OF VEHICLES REGISTERED ON JULY 1 OF THAT CY

Date	Number Sold	Cumulative VMT	Change Since Prev. Jan. 1	Change ÷ N Sold by Prev. July 1
October 1, N-1	0	0		
January 1, N	2,500	4,734,980		
April 1, N	5,000	18,943,711		
July 1, N	7,500	42,626,191		
October 1, N	10,000	75,782,421		
January 1, N+1	10,000	113,588,082	108,853,102	14,514
January 1, N+2	10,000	263,013,854	149,425,772	14,943
January 1, N+3	10,000	409,355,577	146,341,723	14,634
January 1, N+4	10,000	552,336,377	142,980,800	14,298
January 1, N+5	10,000	691,722,187	139,385,810	13,939
January 1, N+6	10,000	827,294,882	135,572,695	13,557
January 1, N+7	10,000	958,877,898	131,583,016	13,158
January 1, N+8	10,000	1,086,319,671	127,441,773	12,744
January 1, N+9	10,000	1,209,500,199	123,180,528	12,318

Vehicles began to sell on October 1 of CY N-1. By January 1 of CY N, 2,500 vehicles had been sold and they had accumulated 4,734,980 VMT. On July 1 of CY N, the vehicles will make their first appearance on the NVPP file, which will say that 7,500 of them had been sold and registered.[62] By January 1, N+1, all 10,000 had been sold and on the road for at least 3 months, and cumulative VMT were 113,588,082. Thus, the 10,000 vehicles of MY N accumulated 108,853,102 VMT during the full CY N (from January 1, N through December 31, N). Because 7,500 had been registered on July 1, N, the VMT during the full CY N divided by registrations

---

[62] In fact, for all MY 2002-2006 cars and LTVs on the NVPP files, 58,442,725 were registered on July 1 of CY=MY and 78,081,646 on July 1 of CY=MY+1; 58,442,725/78,081,646 = 74.8%, nearly identical to the 75% assumed in the Table 2-3 simulation.

on July 1, N is 14,514. In other words, for a make-model with typical mileage, the VMT for model year N in the full calendar year N may be estimated by multiplying the NVPP registration count on July 1, N by 14,514. Table 2-3 also shows that the VMT for model year N in the full calendar year N+1 may be estimated by multiplying the NVPP registration count on July 1, N+1 by 14,943; in CY N+2, by multiplying the NVPP registration count on July 1, N+2 by 14,634; and so on.

Table 2-3 estimates the VMT of the average vehicle of a specific MY during a specific CY. But, of course, some makes and models are driven more than average and others less. R.L. Polk has recently assembled a large vehicle database, derived primarily from repair orders at dealerships and other repair facilities, that includes the vehicle's odometer reading on that day – e.g., on the day it was brought in for maintenance or repairs. In October 2010, Polk extracted records of the MY 2000-2007 cars and LTVs that appear on the database and had at least one odometer reading during the past 30 months. For each individual vehicle, they selected the single, most recent odometer reading for that vehicle in the database (nearly 38% of the cars and LTVs registered in the United States also appear on this database). They grouped the vehicles by the NVPP coding system for makes and models, indicating the number of vehicles and average odometer reading for each code. NHTSA purchased the aggregated file. The agency further aggregated the data by its 5-digit vehicle-group and make-model codes, by body type, and by MY, applying the same programs as in the NVPP data reduction.

The average odometer reading for all MY 2004 cars and LTVs on this Polk file might be, for example, 50,000 (the "typical" vehicle). The average odometer reading for a 2004 _____ 4-door sedan might be 40,000, which would be 80 percent of the reading for the typical vehicle. Thus, in each CY, the VMT for the 2004 _____ 4-door sedan will be an estimated 80 percent of the VMT for the typical vehicle. For example, Table 2-3 shows that the typical MY 2004 vehicle was driven 14,943 miles in CY 2005; the 2004 _____ 4-door sedan was driven an estimated 11,954 miles in CY 2005. Similarly, Table 2-3 shows 14,634 miles for the typical MY 2004 vehicle in CY 2006; that would be 11,707 for the _____ 4-door sedan.

The estimated total VMT of a specific make-model of model year MY in calendar year CY is the number of vehicles registered for that MY and CY times the VMT of the typical vehicle of that MY in that CY, as shown in Table 2-3, times the ratio of odometer readings for that specific vehicle in that MY relative to the typical vehicle.

The file of odometer readings is not a historical database. For any vehicle, even back to MY 2000, it only specifies the most recent reading during the 30 months before October 2010, rather than a series of readings from 2000 onward. The assumption in the above estimates is that if the 2004 _____ 4-door sedan currently has 80 percent of the VMT of the typical MY 2004 vehicle, it also accumulated VMT, year by year, at 80 percent of the typical rate. The database will not identify make-models whose lifetime VMT is concentrated more than usual in the early or the late years of their time on the road.

2.5 Induced-exposure crashes: State data reduction

The preceding data estimates the VMT accumulated by vehicles of a specific curb weight in a given MY and CY but say nothing about who was driving the vehicles, or on what type of road.

Classification of the VMT by age, gender, urban/rural, etc. allow fatality rates to be adjusted for these control variables – i.e., to compare the fatality rates of cars of two different curb weights for drivers of the same age and gender on the same type of road. State data on primarily nonfatal[63] crashes, specifically, "induced-exposure"[64] crash involvements, supply this information. Induced-exposure crash involvements are the non-culpable vehicles in two-vehicle collisions. Those non-culpable vehicles did nothing to precipitate the collision, but were hit merely because "they were there." The involvements are a surrogate for exposure, because they measure how often vehicles "were there" to be hit by other vehicles. "The induced exposure concept assumes that the not-at-fault driver in a two-vehicle crash is reflective of what is 'on the road' at that point in time, and that the sample of all not-at-fault drivers can be used to predict the characteristics of all non-accident involved drivers on the roadway (i.e., exposure characteristics)."[65]

As of April 2012, NHTSA had access to 13 State files through CY 2008 or at least 2007 with relatively complete data on the VINs of crash-involved vehicles:

Alabama	Florida	Kansas	Kentucky
Maryland	Michigan	Missouri	Nebraska
New Jersey	Pennsylvania	Washington State	Wisconsin
Wyoming			

Maryland, Michigan, Nebraska, New Jersey, Pennsylvania, Washington State, and Wisconsin have lower-than-national-average fatality risk, as defined in Section 2.2, while Alabama, Florida, Kansas, Kentucky, Missouri, and Wyoming are higher than average.[66]

Records of induced-exposure crash involvements of MY 2000-2007 cars and LTVs with decodable VINs are extracted. These are limited to vehicles that were in crashes involving exactly two vehicles and zero non-occupants. They are further limited to crashes where one vehicle can be classified "non-culpable" and the other "culpable." If so, the non-culpable vehicle becomes the induced-exposure case. The first criterion for culpability is whether the State itself notes a violation or a contributing circumstance for the driver of that vehicle. If, by

---

[63] The file of induced-exposure crash involvements includes fatal as well as nonfatal crashes; however, over 99 percent of the crashes are nonfatal.

[64] Stutts, J. C., and Martell, C. (1992), "Older Driver Population and Crash Involvement Trends, 1974-1988," *Accident Analysis and Prevention*, Vol. 28, pp. 317-327; Haight, F.A. (1970), "A Crude Framework for Bypassing Exposure," *Journal of Safety Research*, Vol. 2, pp. 26-29; Haight, F.A. (1973). "Induced Exposure," *Accident Analysis and Prevention*, Vol. 5, pp. 111-126; Thorpe, J.D. (1964), "Calculating Relative Involvement Rates in Accidents without Determining Exposure," *Australian Road Research*, Vol. 2, pp. 25-36; Van Der Zwaag, D.D. (1971), "Induced Exposure as a Tool to Determine Passenger Car and Truck Involvement in Accidents," *HIT Lab Reports*, Vol. 1, pp. 1-8; Cerrelli, E. (1973). "Driver Exposure: The Indirect Approach for Obtaining Relative Measures," *Accident Analysis and Prevention*, Vol. 5, pp. 147-156.

[65] Stutts and Martell (1992), p. 318; however, see also Kahane (2003), pp. 34-35 for some caveats about induced-exposure data.

[66] Four State files were included in NHTSA's 2003 and 2010 analyses but not here: Ohio stopped reporting VINs in the data it sends to NHTSA; North Carolina and Utah have not sent 2007 data to NHTSA; Illinois does not report speed limits in the data it sends to NHTSA and the 2003/2010 attempts to impute speed limits from other variables were not fully satisfactory. But in the meantime Michigan has resumed reporting VINs and eight additional States send files including VINs and speed limits to NHTSA: Alabama, Kansas, Kentucky, Nebraska, New Jersey, Washington, Wisconsin, and Wyoming.

that criterion, both vehicles are culpable, the case is not used. But if, by that initial criterion, neither vehicle is culpable in over 20 percent of the crashes, two additional criteria are considered, as in NHTSA's previous reports:[67] (1) if one moving vehicle hit another that was standing still (where permitted), the moving vehicle is culpable; (2) for States coding the impact area, if one vehicle frontally impacts the rear of another vehicle (that was not backing up or encroaching into the first vehicle's lane), the frontally impacting vehicle is culpable. The culpable vehicle may be any type or model year; only the non-culpable vehicle has to be a MY 2000-2007 car or LTV. Furthermore, the non-culpable vehicle must have a driver age 14-96, thereby automatically excluding unoccupied, parked vehicles from the study. In the 13 States, the proportion of two-vehicle crashes with exactly one non-culpable vehicle, by these criteria, ranged from 81 to 97 percent.

Control variables are defined for induced-exposure vehicles parallel to those defined in FARS. DRVAGE, DRVMALE, RURAL (based on the county's population density), CY, and HIFAT_ST can be defined in each State just as in FARS. Many States, like FARS, have a single speed limit defined at the crash level, but some define a speed limit for each vehicle, as vehicles on different roads could collide at an intersection. In these States, for consistency with FARS and the other States, the higher speed limit of the two vehicles is used to define SPDLIM55. All States except Alabama in 2002-2006 specify the time of the crash, permitting NITE to be defined as in FARS. For 2002-2006 Alabama data (which define neither the time of day nor the month of the crash), NITE is imputed from the case number and the light condition, based on the relationship of those variables in the 2007 data (which does encode time and month). For example, the crashes with case numbers near the middle of the file occurred in June and July, at which time virtually all crashes in the dark are between 7:00 P.M. and 5:59 A.M. (NITE = 1), whereas the crashes with case numbers near the end of the file occurred in December, at which time virtually all daylight crashes are between 6:00 A.M. and 6:59 P.M. (NITE = 0). This is the only imputed control variable, and only in 2002-2006 Alabama data.

As of April 2012, 12 of the 13 State files were available at NHTSA through 2008; only one of the smallest States did not yet have 2008 data available. This is a change from NHTSA's 2011 preliminary report, when 7 of the 13 State files were available at NHTSA only through 2007.[68] For the one State where CY 2008 data is still not available, the CY 2007 induced-exposure data is used to classify the CY 2008 as well as the CY 2007 vehicle years and VMT by driver age and gender, urban/rural, etc. The assumption here is that the distribution of those variables would not be likely to change much in one year. (The preliminary report followed that procedure for 7 States.) In addition, 2002 Pennsylvania data are not available at NHTSA and 2002-2003 Michigan data have few or no VINs. The 2003 Pennsylvania are used to classify the CY 2002 vehicle years and VMT, while the 2004 Michigan data are used to classify both CY 2002 and 2003.

The counts of induced-exposure crash involvements in 2002-2008 (including duplicate cases that fill in the missing calendar years) vary from State to State:

---

[67] Kahane (2003), p. 32.
[68] Kahane (2011), p. 47.

Alabama	225,716
Florida	337,754
Kansas	61,897
Kentucky	189,586
Maryland	123,956
Michigan	382,981
Missouri	223,868
Nebraska	53,972
New Jersey	416,711
Pennsylvania	128,693
Washington State	156,848
Wisconsin	141,227
Wyoming	14,019
	2,457,228

There are usually more crashes in the more populous States, but reporting thresholds play a role. For example, Pennsylvania has considerably fewer crashes than New Jersey because of a higher reporting threshold. The technique described in the next section, weighting the cases by VMT, will give higher weights to the cases in the States with higher reporting thresholds.

This report relies on induced-exposure data from 13 States to represent the United States. Although the absolute distributions of crashes by driver age, rural/urban, etc. differ considerably from State to State, the interactions of these variables with curb weight and footprint are remarkably consistent across States. Section 3.5 will show that the use of data from just 13 States rather than all the States makes minimal-to-moderate contribution to the uncertainty of the estimated effects of mass reduction on fatality rates.

2.6 Assembling the analysis data files

The database of induced-exposure crash involvements for model year 2000-2007 vehicles in calendar years 2002-2008 consists of 2,457,228 records from 13 States, as described in the preceding section. The critical step in building the database is to allocate the right number of vehicle years and VMT to each induced-exposure crash, so that the induced-exposure crashes in 13 States represent all the vehicle years and VMT in the United States. This will apportion the nation's vehicle registration years and VMT not only by make-model, body style, model year and calendar year but also by driver age, gender, rural/urban location and the other control variables. The weighting procedure has been modified somewhat from NHTSA's 2003 and 2010 reports to allow greater flexibility in addressing vehicles with lower sales that might not experience induced-exposure crashes every year in each State, especially in this study, where some of the 13 States have relatively low populations and few crashes.

The first example of allocation is for a vehicle with high sales. During CY 2007, the MY 2005 Toyota Camry 4-door had the following non-zero registrations (as of July 1, 2007) and non-zero counts of induced-exposure crash involvements in each of the 13 States:

MY 2005 Toyota Camry in CY 2007	Registrations	Induced-Exposure Involvements
Alabama	6,811	182
Florida	34,548	248
Kansas	2,973	29
Kentucky	4,934	125
Missouri	5,242	116
Wyoming	323	5
These 6 high-fatality-rate States	54,831	
Maryland	10,952	97
Michigan	4,664	81
Nebraska	1,361	19
New Jersey	15,829	315
Pennsylvania	13,543	58
Washington State	7,030	99
Wisconsin	5,508	63
These 7 low-fatality-rate States	58,887	
All 24 high-fatality-rate States + D.C.	158,735	
All 26 low-fatality-rate States	233,269	
Entire United States	392,004	

Since there were 182 crash involvements and 6,811 registered cars in Alabama, each crash corresponds to

$$6,811/182 = 37.42 \text{ vehicle years within Alabama}$$

However, since the 6 high-fatality-rate States in our sample had 54,831 registered vehicles, whereas all 24 high-fatality-rate States plus D.C. had 158,735 registered vehicles, each Alabama crash is allocated

$$(158,735/54,831) \times (6,811/182) = 108.34 \text{ high-fatality-rate vehicle years in the United States}$$

Similarly, each of the 97 crash involvements in Maryland is allocated

$$(233,269/58,887) \times (10,952/97) = 447.26 \text{ low-fatality-rate vehicle years in the United States}$$

The allocation of vehicle years per crash in the 13 States is:

MY 2005 Toyota Camry in CY 2007	Induced-Exposure Involvements	Vehicle Years Apportioned Per Involvement
Alabama	182	108.34
Florida	248	403.29
Kansas	29	296.79
Kentucky	125	114.27
Missouri	116	130.82
Wyoming	5	187.02
Maryland	97	447.26
Michigan	81	228.09
Nebraska	19	283.75
New Jersey	315	199.06
Pennsylvania	58	924.96
Washington State	99	281.29
Wisconsin	63	346.33

Note that

$$182 \times 108.34 + 248 \times 403.29 + 29 \times 296.79 + 125 \times 114.27 + 116 \times 130.82 + 5 \times 187.02$$

$$+ 97 \times 447.26 + 81 \times 228.09 + 19 \times 283.75 + 315 \times 199.06 + 58 \times 924.96 + 99 \times 281.29 + 63 \times 346.33$$

$$= 392{,}004 \text{ vehicle years in the entire United States}$$

In other words, these weight factors (vehicle years) allocated to each induced-exposure crash will add up, over the entire file, exactly to the number of 2005 Toyota Camry 4-door registered in the United States during CY 2007. (In general, the weight factors are higher in States such as Pennsylvania that have higher crash-reporting thresholds, and relatively fewer reported crashes per vehicle year.)

The second example is the MY 2004 BMW 740iL, a vehicle with lower sales that did not experience an induced-exposure crash in several States during CY 2007:

MY 2004 BMW 740iL in CY 2007	Registrations	Induced-Exposure Involvements
Alabama	124	1
Florida	1,349	11
Kansas	26	none
Kentucky	52	1
Missouri	82	none
Wyoming	1	none
The 3 high-fatality-rate States with registrations and crashes	1,525	
Maryland	283	3
Michigan	177	1
Nebraska	11	none
New Jersey	527	9
Pennsylvania	313	none
Washington State	77	1
Wisconsin	60	none
The 4 low-fatality-rate States with registrations and crashes	1,064	
All 24 high-fatality-rate States + D.C.	4,004	
All 26 low-fatality-rate States	6,513	
Entire United States	10,517	

Since there was 1 crash involvement and 124 registered cars in Alabama, that crash corresponds to 124 vehicle years within Alabama

However, since the 3 high-fatality-rate States <u>with non-zero registrations and crashes</u> in our sample had 1,525 registered vehicles, whereas all 24 high-fatality-rate States plus D.C. had 4,004 registered vehicles, each Alabama crash is allocated

$(4,004/1,525) \times (124/1) = 325.57$ high-fatality-rate vehicle years in the United States

The allocation of vehicle years per crash in the 13 States is:

MY 2004 BMW 740iL in CY 2007	Induced-Exposure Involvements	Vehicle Years Apportioned Per Involvement
Alabama	1	325.57
Florida	11	321.99
Kansas	none	N/A
Kentucky	1	136.53
Missouri	none	N/A
Wyoming	none	N/A
Maryland	3	577.44
Michigan	1	1083.46
Nebraska	none	N/A
New Jersey	9	358.43
Pennsylvania	none	N/A
Washington State	1	471.34
Wisconsin	none	N/A

Again, 1x325.57 + 11x321.99 + 1x136.53 + 3x577.44 + 1x1083.46 + 9x358.43 + 1x471.34

= 10,517 vehicle years in the entire United States

This process is repeated for all other make-models of cars and LTVs, MY 2000-2007 in CY 2002-2008. It successfully allocates 99.81 percent of the registration years of MY 2000-2007 vehicles in CY 2002-2008. However, some vehicle group-make-model-body style-MY combinations had sales so low that they had zero crashes and/or zero registrations in all 6 high-fatality-rate States and/or in all 7 low-fatality-rate States in one or more CY, making it impossible to directly allocate registration years to a specific crash in that CY. For such combinations, the induced-exposure cases for the various other CY are accepted (but only the registrations for that CY are allocated among them). If the sales are so low that none of the other CY produced candidate cases for that vehicle group-make-model-body style-MY combination, even the induced-exposure cases for the various other MY, but same vehicle group, make-model, and body style are accepted. If, even with this extension, there are still no candidate cases, this combination is deleted from the working induced-exposure file and likewise from the file of fatal crash involvements. As a consequence, combinations totaling only 0.01 percent of the registration years of MY 2000-2007 vehicles in CY 2002-2008 and only 4 of the over 113,000 fatality cases were deleted from the analysis. The new procedure eliminates the unappealing technique in the 2003 and 2010 reports of creating single dummy records that average, over several crashes, the values of fundamentally categorical variables such as the driver's gender.[69]

VMT are also allocated to each induced-exposure case, based on the make-model, body style, MY, and CY (see Section 2.4). The average 2005 Toyota Camry 4-door is estimated to have

---

[69] Kahane (2003), p. 37; however, changing the procedure likely had little influence on the results, because dummy records accounted for only 1% of the registration years in the previous study and cases from alternate CY/MY accounted for only 0.19% of the registration years in this report.

traveled 12,196 miles in CY 2007. Since each Alabama crash is apportioned 108.34 vehicle years, it is also apportioned 108.34 x 12,196 = 1,321,315 vehicle miles. The average 2004 BMW 740iL (a higher-mileage vehicle) is estimated to have traveled 15,886 miles in CY 2007. Since the Alabama crash is apportioned 325.57 vehicle years, it is also apportioned 325.57 x 15,886 = 5,172,005 vehicle miles.

Here are hypothetical examples of a record from the database of fatal-crash involvements and a record from the database of induced-exposure crash involvements. The fatal-crash record is from a high-fatality-rate State and the induced-exposure record is from Alabama, a high-fatality-rate State. Both records are MY 2005 Toyota Camry 4-door sedans in CY 2007:

	Record from Fatal-Crash Database	Record from Induced-Exposure Database
Crash type	Hit truck-based LTV ≥ 4,150 lbs.	-
N of fatalities in the crash	2	-
Vehicle registration years	-	108.34
VMT	-	1,321,315
Vehicle type	4-door car	4-door car
Curb weight	3,248 lbs.	3,248 lbs.
Footprint	45.0 sq. ft.	45.0 sq. ft.
Driver age	24	28
Driver male?	1	1
At night?	0	0
Rural?	1	0
Speed limit 55+?	1	0
Calendar year	2007	2007
Vehicle age	2	2
High-fatality-rate State?	1	1
ABS (4-wheel)	1	1
ESC	.07	.07
AWD	0	0
Curtain air bags	0	1
Rollover curtains	0	0
Torso air bags	0	1
Combination side air bags	0	0
IIHS overall frontal rating	1 (good)	1 (good)

The 113,248 records on the database of fatal crash involvements come from all 50 States and the District of Columbia. Each of the 2,457,228 records on the database of induced-exposure crash involvements is nominally a specific crash involvement in one of 13 States, a discrete unit. But when any of these records is weighted by its allocation of vehicle years or VMT, it becomes a cohort of vehicle years or VMT in the United States. The induced-exposure records are a national census of vehicle years and VMT. Fatal-crash records are weighted by the number of fatalities in the crash. The sum of the fatalities in the crashes divided by the sum of the VMT is the national fatality rate per mile. The two databases will be combined and used for the regression analyses.

3. Results

3.0 Summary

The effects of mass reduction and footprint reduction on societal fatality rates per billion VMT are estimated by logistic regression for five classes of vehicles in nine types of crashes, controlling for other factors such as driver age and gender, urbanization and speed limit, and safety equipment such as ESC. The regressions tend to show a benefit for mass reduction in rollovers and fixed-object impacts. They tend to show a harmful effect in collisions with other vehicles, except for a benefit for mass reduction in a heavy vehicle in collisions with lighter vehicles. Averaged across crash types, mass reduction by 100 pounds in cars weighing less than 3,106 pounds is associated with a statistically significant 1.56 percent societal fatality increase (confidence bounds, 0.39 to 2.73 percent); in cars weighing 3,106 pounds or more and in truck-based LTVs weighing less than 4,594 pounds, smaller, non-significant increases; and in LTVs weighing 4,594 pounds or more and in CUVs/minivans, non-significant benefits. The report presents several computational examples of simultaneous mass reductions in all types of vehicles: (1) mass reduction by 100 pounds in every vehicle would not significantly increase fatalities; (2) mass reductions that are proportionately somewhat greater in the heavier vehicles can be safety-neutral as point estimates. However, the estimated safety effects that will appear in the PRIA will be based on the CAFE Compliance and Effects Modeling System (usually referred to as the "Volpe model") and not these computational examples.

This chapter's analyses largely recapitulate Chapter 3 of NHTSA's November 2011 preliminary report, except they use the updated database that includes CY 2008 data for six additional States and a revised estimate of annual VMT. Results are quite similar to the preliminary report. The results are also directionally identical to NHTSA's 2010 report but the fatality increase in the lighter cars as well as the reduction in the heavier LTVs both appear to be of smaller magnitude than in the earlier study. Chapter 4 contains new analyses addressing peer-review and public comments.

3.1 Regression setup

Case-vehicle categories: NHTSA's 2003 and 2010 reports analyzed two groups of MY 1991-1999 case vehicles, passenger cars and LTVs. All LTVs were included in the same analysis, with categorical variables to indicate the LTV type (SUV, MINIVAN, BIGVAN; pickup trucks being the default type). NHTSA expressed misgivings, echoed by Paul Green in his peer review of the 2010 report, that including disparate types of "niche" vehicles, each with its own pattern of crash types and of relationships between mass and footprint, might generate coefficients for mass and footprint that reflect the vehicle mix rather than the underlying relationships, within each individual type of LTV, of these parameters with fatality risk.[70] The issue is now even more critical because of the increase in CUVs, which are technically LTVs but in many ways more closely resemble cars. This report, like the November 2011 preliminary report will present regression analyses on three rather than two classes of vehicles, in an attempt to mitigate this issue:

---

[70] Kahane (2010), pp. 522-523.

- Truck-based LTVs will exclude CUVs and, for that matter, minivans, which also resemble CUVs and cars in some ways. CUVs and minivans are examined separately in their own group. In order to make this class even more homogeneous, full-size vans, which account for only 1.34 percent of the vehicle registration years and 1.63 percent of the VMT in the database, are also excluded in the regression analyses. In other words, it is limited to pickup trucks and truck-based SUVs, most of which are built on pickup-truck platforms.[71]
- CUVs and minivans: An argument could be made for including them with cars, but they are not really cars; it would just move the problem in the 2003/2010 reports from one class to another. They will constitute a class by themselves. However, in the regressions, minivans will not include the Chevrolet Astro and GMC Safari (which more closely resemble full-size vans than the typical minivan). Appendix A lists the SUVs that this report calls CUVs.
- Passenger cars: As in the 2010 report, 2-door and 4-door cars are included in the regressions. Two-door high-performance cars and 4-door police cars are excluded because their inclusion was found to skew the regression results in the 2010 and 2003 analyses, respectively.[72] To make this class more homogeneous and simplify the analyses, the 3.5 percent of passenger cars equipped with all-wheel-drive are also excluded (as will be discussed at the end of this section). But all other 2-door and 4-door cars are included. (A sensitivity test at the end of Section 3.5 will include the full-size vans among the truck-based LTVs, and high-performance cars, police cars, and AWD-equipped cars among the passenger cars.)

Formulation of the independent variables: The two principal independent variables are curb weight and footprint. They were highly correlated in MY 1991-1999 and, as the 2010 report discussed at length, that raised concerns about the accuracy of regression coefficients.[73] The correlation coefficient of curb weight with footprint was .893 in cars and .742 in LTVs. After controlling for body style, but not the other control variables, the variance inflation factors (VIF) were 6.1 in cars and ranged from 3.6 to 5.7 in LTVs. In logistic regressions, "there is no formal cutoff value to use with VIF for determining presence of multicollinearity. Values of VIF exceeding 10 are often regarded as indicating multicollinearity, but in weaker models, which is often the case in logistic regression, values above 2.5 may be a cause for concern."[74] Allison "begins to get concerned" when he sees VIF scores over 2.5.[75]

---

[71] Although full-sized vans and the Chevrolet Astro/GMC Safari are not included in the regression analyses, these vehicles are included in the tabulations of baseline fatalities of truck-based LTVs and regression coefficients for truck-based LTVs are applied to those baseline fatalities.

[72] *Ibid.*, pp. 484-486; Kahane (2003), pp. 171-172; in this study, the excluded high-performance cars are Chrysler/Plymouth Prowler, Dodge Viper, Ford Mustang and GT, Chevrolet Corvette and Camaro, Pontiac GTO and Firebird, BMW Z3 and Z8, Jaguar XK, Mercedes SL and SLR, all Porsche, and Acura NSX. Excluded 4-door cars are not limited to "police" models of Ford Crown Victoria and 2006-2007 Chevrolet Impala but also all other Crown Victorias (which, if not police cars, are often taxicabs or high-mileage fleet vehicles) and 2004-2005 Impala SS, which often served as a police car before Chevrolet developed its "police" model in 2006.

[73] Kahane (2010), pp. 479-481 and 522-523.

[74] Schadler, A. *Multicollinearity in Logistic Regression*. Lexington, KY: University of Kentucky Center for Statistical Computing Support. http://www.uky.edu/ComputingCenter/SSTARS/MulticollinearityinLogisticRegression.htm.

[75] Allison, P.D. (1999), pp. 48-51.

These statistics have changed little in the MY 2000-2007 database. The correlation coefficient of curb weight with footprint is now .896 in cars, .748 in truck-based LTVs, and .781 in CUVs and minivans. After controlling for body style <u>and</u> all the other control variables, the VIF is 7.34 in cars, 9.80 in truck-based LTVs, and 8.71 in CUVs and minivans.[76] VIF continues to be in the somewhat unfavorable 2.5-10 range.

Several methods to index curb weight to footprint or vice-versa were considered in response to comments by the peer reviewers on the 2010 and also the 2011 preliminary report. The analyses will be presented in Section 4.5, which will show that the indexing techniques successfully lower VIF but do not meaningfully affect the results of the logistic regressions. NHTSA will continue to use curb weight and footprint as the two principal variables, but will check the results with a decile analysis – by a method revised in response to Farmer's peer review of the 2010 report – in Section 3.4. Unlike the 2010 report, the revised decile analysis will produce results consistent with the basic regressions and provide at least some corroboration for the accuracy of the coefficients.

Footprint is measured in square feet. As in the 2003 and 2010 reports, curb weight is entered as a 2-piece linear variable (measured in hundreds of pounds) in the analyses of cars and truck-based LTVs, to permit separate estimates of the effects of mass reduction in the lighter or in the heavier vehicles. However, because vehicles became heavier after 2000 (and because CUVs and minivans are no longer included with the truck-based LTVs), the median weight of MY 2000-2007 vehicles involved in fatal crashes, which serves as the dividing line between "lighter" and "heavier," has increased in the cars from 2,950 to 3,106 and from 3,870 (including CUVs and minivans) to 4,594 (excluding them) in the LTVs. In the analyses of passenger cars, for example, if the curb weight is less than 3,106,

$$UNDRWT00 = .01 \text{ (curb weight} - 3,106), OVERWT00 = 0$$

And if it is 3,106 or more,

$$UNDRWT00 = 0, OVERWT00 = .01 \text{ (curb weight} - 3,106)$$

In the regression analyses of CUVs and minivans, where the database is much smaller as shown in Table 2-1, it is futile to estimate separate effects above and below the median weight. Mass is a simple linear variable, LBS00, measured in hundreds of pounds.

Dichotomous variables indicate the vehicle's body style. In the regressions of passenger cars, TWODOOR=1 for 2-door cars, 0 for 4-door cars. In the regressions of truck-based LTVs, SUV=1 for SUVs, HD_PKP=1 for heavy-duty pickup trucks (200/300 series), and both variables are zero for 100-series and smaller pickup trucks. In the regressions of CUVs and minivans, MINIVAN=1 for minivans, 0 for CUVs.

---

[76] As the VIF test is based on a linear regression model with a dummy dependent variable, control variables such as driver age and calendar year are formulated as simple linear variables. By contrast, the VIF for curb weight with all the other control variables, but not footprint, is only 2.43 for cars, 1.86 for truck-based LTVs, and 2.83 for CUVs and minivans.

Fatal-crash rates per VMT are higher for young and old drivers than for people age 30-50. They are higher for males than females, but the difference decreases at the higher ages. As in the 2003 and 2010 reports, these relationships are captured by entering driver age and gender as nine variables, DRVMALE (defined in Section 2.2) and M14_30, M30_50, M50_70, M70_96, F14_30, F30_50, F50_70, F70_96, where, for example, M14_30 = 30 − DRVAGE for male drivers age 14-30, = 0 for male drivers age 31+ and all female drivers.[77] In other words, age is entered as a 4-piece linear variable, allowing separate slopes for various age groups depending on the type of crash and the driver's gender.

The control variables NITE, RURAL, SPDLIM55, and HIFAT_ST; the vehicle attributes ESC, ABS, and AWD; and the side-air-bag variables CURTAIN, ROLLCURT, TORSO, and COMBO are unchanged from Section 2.2. Information on compatibility certification is expressed as two dichotomous variables, BLOCKER1=1 if the LTV is certified to meet Option 1, BLOCKER2=1 if Option 2. As in the 2003 and 2010 reports, vehicle age is expressed by VEHAGE = CY - MY, with an additional variable BRANDNEW = 1 if VEHAGE = 0 (to allow an effect for new vehicles that differs from the trend). Calendar year is expressed by the dichotomous variables CY2002, CY2003, CY2004, CY2005, CY2007, and CY2008, all of which equal zero if CY = 2006.

IIHS has not crash-tested every make-model in the database. The test results are not included among the variables in the basic regression analyses, but in a separate group of analyses, limited to the tested make-models, to be discussed in Section 3.8

Setup for a regression: As explained in Section 2.6, a regression is performed on a temporary data file that combines a subset of the records from the database of fatal crash involvements and a subset of the records from the database of induced-exposure crash involvements. For example, the analysis of first-event rollovers of passenger cars would combine the subset of fatal-involvement records that are passenger cars (vehicle class) and first-event rollovers (fatal-crash type) with the subset of induced-exposure records that are passenger cars (vehicle class). The two databases already have many variables in common: curb weight, footprint, and the various controls such as driver age and gender that will be independent variables in the regression. One new variable, the dependent variable FATAL, is defined on the temporary, combined data file: FATAL equals 1 for each of the records from the database of fatal-crash involvements and FATAL equals 2 for each of the records from the database of induced-exposure involvements. The other new variable is the case-weight factor for the regression, WEIGHTFA. Each record from the database of fatal-crash involvements is weighted by the number of fatalities in the crash, including occupants of other vehicles and non-occupants (WEIGHTFA equals the FARS variable FATALS). Each record from the database of induced-exposure involvements is weighted by its allocation of the nation's VMT (WEIGHTFA equals the variable VMTWTFA on the induced-exposure database). The technique is logistic regression performed by the SAS® procedure LOGISTIC. The regressions analyze rates of fatal-crash <u>involvements</u> (weighted by the number of fatalities in the crash); the computational models that will be presented in Section 3.6 translate the regression results into estimates of the effect of mass reduction on overall <u>fatalities</u>.

---

[77] Kahane (2003), pp. 69-70.

Exploratory analyses (variables that may be dropped): In NHTSA's 2003 report, the analysis of passenger cars was simplified because: (1) the average curb weight of passenger cars changed little from MY 1991 to 1999; (2) according to the data used in that study, lighter and heavier cars had about the same annual VMT. That made it possible to analyze fatality rates per car registration year rather than per VMT and to drop the vehicle-age and CY variables. Neither assumption holds for MY 2000-2007. Exploratory regressions omitting some of the control variables showed that only a few variables may be dropped from any of the analyses without perceptibly affecting the results. The variables for the various types of side air bags may be dropped from the analyses of pedestrian crashes, because the air bags in the vehicle will not affect the fatality risk for the pedestrians. In rollover crashes, ROLLCURT is the only air-bag variable; in the other non-pedestrian crash types, ROLLCURT is omitted but CURTAIN, TORSO, and COMBO are included. However, when the case vehicles are truck-based LTVs, all the air-bag variables as well as ABS may be omitted for all crash types: not because these technologies are ineffective, but because they are about equally available in lighter and heavier LTVs (whereas in cars, the heavier vehicles are more likely equipped with side air bags and ABS). The blocker-beam variables, on the other hand appear only in the regressions for the truck-based LTVs.

In general, the variable for all-wheel/4-wheel drive (AWD) should not be dropped from the regression analyses, because the larger and heavier vehicles tend to be equipped somewhat more often with those technologies. However, only 3.5 percent of passenger cars were equipped with AWD in MY 2000-2007. The analyses of passenger cars are simplified by limiting them to cars without AWD and dropping the AWD control variable.

3.2 Three regression examples

There are 27 basic regressions: three vehicle classes x nine crash types. Here, for example, is the regression for passenger cars' collisions with truck-based LTVs weighing 4,150 pounds or more. This is the regression that associated the largest fatality increase with lower mass. The independent variables are the mass and footprint of the passenger cars (case vehicles). The mass of the LTV (partner vehicle) is not specified in the regression, only that it weighs 4,150 pounds or more. The temporary data file for the analysis consists of 1,245,705 records, including 3,622 from the fatal-crash database and 1,242,083 from the induced-exposure database. In the 3.91 trillion VMT allocated among the 1,242,083 induced-exposure crash involvements of MY 2002-2007 passenger-car case vehicles during CY 2002-2008 (see "Response Profile"), these cars experienced 3,622 fatal collisions with truck-based LTVs ≥ 4,150 pounds, resulting in 4,256 fatalities (most of whom were occupants of the cars but some were occupants of the LTVs):

```
The LOGISTIC Procedure: Car collisions with LTVs ≥ 4,150 pounds

          Response Profile

 Ordered                Total          Total
   Value    FATAL   Frequency         Weight

       1        1        3622           4256
       2        2     1242083   3.9063533E12
```

Model Fit Statistics

Criterion	Intercept Only	Intercept and Covariates
AIC	184180.74	175313.61
SC	184192.77	175686.70
-2 Log L	184178.74	175251.61

R-Square 0.0071   Max-rescaled R-Square   0.0520

Testing Global Null Hypothesis: BETA=0

Test	Chi-Square	DF	Pr > ChiSq
Likelihood Ratio	8927.1255	30	<.0001
Score	12633.6468	30	<.0001
Wald	9700.9776	30	<.0001

Analysis of Maximum Likelihood Estimates

Parameter	DF	Estimate	Standard Error	Wald Chi-Square	Pr > ChiSq
Intercept	1	-22.0429	0.5795	1447.0531	<.0001
UNDRWT00	1	-0.0606	0.0111	29.7995	<.0001
OVERWT00	1	-0.0234	0.0133	3.1064	0.0780
FOOTPRNT	1	-0.0177	0.0129	1.8852	0.1697
TWODOOR	1	-0.0247	0.0463	0.2834	0.5945
CURTAIN	1	-0.0697	0.0681	1.0481	0.3059
COMBO	1	-0.1583	0.0619	6.5495	0.0105
TORSO	1	-0.1727	0.0588	8.6198	0.0033
ABS	1	0.00945	0.0554	0.0291	0.8646
ESC	1	-0.3159	0.0961	10.8148	0.0010
DRVMALE	1	0.2567	0.0880	8.5031	0.0035
M14_30	1	0.0598	0.00709	71.0182	<.0001
M30_50	1	0.00935	0.00477	3.8356	0.0502
M50_70	1	0.0403	0.00515	61.3510	<.0001
M70_96	1	0.1017	0.00629	260.8459	<.0001
F14_30	1	0.0572	0.00765	55.9328	<.0001
F30_50	1	-0.00831	0.00495	2.8131	0.0935
F50_70	1	0.0584	0.00511	130.5645	<.0001
F70_96	1	0.0899	0.00740	147.5081	<.0001
NITE	1	0.7325	0.0346	449.1182	<.0001
RURAL	1	1.4222	0.0322	1954.7903	<.0001
SPDLIM55	1	1.8474	0.0315	3445.4578	<.0001
HIFAT_ST	1	0.3839	0.0313	150.4042	<.0001
VEHAGE	1	0.0410	0.00992	17.1056	<.0001
BRANDNEW	1	0.1365	0.0583	5.4761	0.0193
CY2002	1	-0.1667	0.0739	5.0887	0.0241
CY2003	1	-0.0474	0.0631	0.5654	0.4521
CY2004	1	-0.0550	0.0579	0.9046	0.3416
CY2005	1	0.00559	0.0533	0.0110	0.9165
CY2007	1	-0.00795	0.0501	0.0251	0.8741
CY2008	1	-0.1568	0.0533	8.6562	0.0033

Societal fatality risk was an estimated 6.06 percent lower for each 100-pound increment across the cars weighing less than 3,106 pounds (see the entry for UNDRWT00 in "Analysis of Maximum Likelihood Estimates"). The regression printouts generated by SAS® indicate the effect of <u>increasing</u> mass by 100 pounds. However, this report, consistent with earlier NHTSA reports will show in all tables of results the effects of <u>reducing</u> mass by 100 pounds. In other words, the coefficients that will be shown in the tables will be the <u>opposites</u> of the coefficients in these regression examples.

The Wald chi-square for that coefficient would have been 29.80 if each fatality and each mile of VMT had been drawn from a simple random sample (SRS), but it was not; a higher sampling error that takes the study design into account is computed in Section 3.6. A Wald chi-square of 3.84 or more indicates statistical significance at the .05 level for SRS. If the SAS® printout shows 3.84 or more, the effect may or may not be significant given the actual sample design, but if it shows less than 3.84 even for SRS, the effect will almost certainly not be significant after the sample design is taken into account. Risk decreased by an estimated 2.34 percent as mass increased by 100 pounds across cars ≥ 3,106 pounds; with a chi-square of 3.11, this would not have been significant even under simple random sampling. Risk also decreased by 1.77 percent as footprint increased by 1 square foot; this, too, is not statistically significant. Two-door and 4-door cars had about the same fatality rate, after controlling for the other variables. The three types of side air bags and ESC are associated with fatality reductions (although the amount of reduction in any individual regression has wide error bounds), but ABS has little effect. Fatality rates increase for males and for each year that a driver is younger than 30 or older than 50 (especially if older than 70). Fatality rates per VMT are substantially higher at night, on rural roads, when the speed limit is 55+, and in high-fatality States. Fatality rates increase with vehicle age, but they are also higher when the car is less than a year old. Rates were lower in the earlier calendar years (when there were fewer heavy-LTV potential partner vehicles on the road) and also in 2008 (when fatalities in almost all types of crashes dropped sharply). However, this is not the most common pattern of the CY coefficients, but somewhat of a special case because the number of heavy LTVs on the road (and their potential for being partner vehicles in crashes) increased during 2002-2008. The more typical pattern, such as in the various types of single-vehicle crashes, is for the CY coefficients to shift from positive in the early years to negative later on, as fatality rates per VMT decreased over time. And in crashes with partner vehicles that became less common during 2002-2008, such as the lighter LTVs, the shift from early positives to zero or negative is even stronger.

The regression results also include overall "model fit statistics." The "max-rescaled R-square," which has approximately the same meaning as the R-square statistic for a linear regression, is a relatively low .0520. Basically, it says that it is difficult to predict from demographic and environmental variables such as driver age, urbanization, and time of day that one specific mile of travel will result in a fatal crash while another will not. On the other hand, "testing global null hypothesis" finds that the likelihood-ratio chi-square for the model is 8927.13 with 30 degrees of freedom, which is statistically significant at the .0001 level. In other words, the control variables, as a group, have strong relationships with fatality risk, even though they are insufficient to predict if a specific mile of travel will be fatal or not. Similarly, in the other 26 regressions, max-rescaled R-square ranged from .0220 to .1055, while the likelihood-ratio chi-square was always significant at the .0001 level. Neither of these statistics sheds much light on how well the regressions are measuring the relationships between fatality risk and mass or footprint; it is primarily the other control variables with much higher chi-squares, such as NITE, RURAL, and SPDLIM55 that are driving these statistics.

Here is an example of a regression where pickups and truck-based SUVs are the case vehicles, specifically their collisions with cars, CUVs or minivans weighing 3,082 pounds or more. The independent variables are the mass and footprint of the truck-based LTVs. The mass of the car/CUV/minivan partner vehicle is not specified, only that it weighs 3,082 pounds or more:

```
The LOGISTIC Procedure: Pickups/truck-based SUV collisions with cars/CUVs/minivans ≥ 3,082 pounds

                    Response Profile

     Ordered                  Total           Total
      Value       FATAL     Frequency         Weight

        1           1          4928            5892
        2           2        703814       2.8419417E12

            Model Fit Statistics

                                  Intercept
                    Intercept        and
    Criterion         Only        Covariates

    AIC             247397.14      239403.24
    SC              247408.61      239758.84
    -2 Log L        247395.14      239341.24

    R-Square    0.0113    Max-rescaled R-Square    0.0383

         Testing Global Null Hypothesis: BETA=0

    Test                  Chi-Square        DF      Pr > ChiSq

    Likelihood Ratio       8053.9054        30        <.0001
    Score                 10194.9220        30        <.0001
    Wald                   8260.3693        30        <.0001

              Analysis of Maximum Likelihood Estimates

                                Standard         Wald
    Parameter     DF   Estimate   Error      Chi-Square    Pr > ChiSq

    Intercept      1   -21.5161   0.2940      5357.5318      <.0001
    UNDRWT00       1    0.00708   0.00621        1.3009      0.2541
    OVERWT00       1    0.0137    0.00484        8.0457      0.0046
    FOOTPRNT       1   -0.00308   0.00452        0.4638      0.4958
    SUV            1    0.0223    0.0579         0.1490      0.6995
    HD_PKP         1    0.1076    0.0560         3.6855      0.0549
    BLOCKER1       1   -0.1023    0.0313        10.7190      0.0011
    BLOCKER2       1   -0.0503    0.0439         1.3114      0.2521
    DRVMALE        1    0.0146    0.0698         0.0440      0.8339
    M14_30         1    0.0365    0.00514       50.5162      <.0001
    M30_50         1    0.0174    0.00287       36.6518      <.0001
    M50_70         1    0.0194    0.00388       25.1464      <.0001
```

M70_96	1	0.0657	0.0100	42.9428	<.0001
F14_30	1	0.0315	0.00827	14.5042	0.0001
F30_50	1	0.00605	0.00461	1.7192	0.1898
F50_70	1	0.0355	0.00728	23.7240	<.0001
F70_96	1	0.0569	0.0253	5.0366	0.0248
ESC	1	-0.2554	0.0649	15.4612	<.0001
AWD	1	-0.1458	0.0339	18.4581	<.0001
NITE	1	0.8638	0.0288	900.9832	<.0001
RURAL	1	1.1546	0.0280	1695.9116	<.0001
SPDLIM55	1	1.6350	0.0269	3694.0636	<.0001
HIFAT_ST	1	0.1980	0.0284	48.5812	<.0001
VEHAGE	1	0.0328	0.00915	12.8202	0.0003
BRANDNEW	1	0.0506	0.0481	1.1109	0.2919
CY2002	1	0.1766	0.0586	9.0968	0.0026
CY2003	1	0.1188	0.0528	5.0562	0.0245
CY2004	1	0.0475	0.0489	0.9445	0.3311
CY2005	1	0.00207	0.0459	0.0020	0.9641
CY2007	1	-0.0623	0.0438	2.0295	0.1543
CY2008	1	-0.1834	0.0470	15.2648	<.0001

The 2.84 trillion VMT of MY 2000-2007 truck-based LTVs were allocated among the 703,814 induced-exposure records. These LTVs experienced 4,928 fatal collisions with cars, CUVs, and minivans ≥ 3,082 pounds, resulting in 5,892 fatalities (mostly in the cars, CUVs, or minivans). Societal fatality risk was a non-significant 0.708 percent higher for each 100-pound increment across the LTVs under 4,594 pounds. Fatality risk increased by 1.37 percent for each additional 100 pounds in the LTVs weighing 4,594 pounds or more; this would have been significant with simple random sampling (chi-square = 8.05). Footprint does not have a significant effect. As for the specific type of LTV, SUVs have about the same risk as light-duty pickups, but heavy-duty pickups have about 11 percent higher risk, all else being equal. Both types of blocker beams, ESC, and AWD/4x4 are associated with reduced fatality risk. The effects of driver age, the environmental variables, and vehicle age are similar to the preceding regression, but the driver's gender has little effect (most of these LTVs are driven by males). Again, risk dropped in CY 2008.

Here is a regression where CUVs or minivans are the case vehicles and the partner vehicles are heavy trucks. The independent variables are the mass (as a single, linear variable LBS100 rather than the separate variables UNDRWT00 and OVERWT00 in the preceding regressions) and footprint of the CUVs or minivans. The mass of the trucks is not specified, but they range from pickup trucks and vans just over 10,000 pounds GVWR to the heaviest combination vehicles.

The LOGISTIC Procedure: CUV/minivan collisions with heavy trucks

Response Profile

Ordered Value	FATAL	Total Frequency	Total Weight
1	1	680	828
2	2	371009	1.219041E12

Model Fit Statistics

Criterion	Intercept Only	Intercept and Covariates
AIC	36616.278	34264.353
SC	36627.103	34599.954
-2 Log L	36614.278	34202.353

R-Square   0.0065   Max-rescaled R-Square   0.0689

Testing Global Null Hypothesis: BETA=0

Test	Chi-Square	DF	Pr > ChiSq
Likelihood Ratio	2411.9242	30	<.0001
Score	3748.1884	30	<.0001
Wald	2381.7525	30	<.0001

Analysis of Maximum Likelihood Estimates

Parameter	DF	Estimate	Standard Error	Wald Chi-Square	Pr > ChiSq
Intercept	1	-20.8588	0.7381	798.6449	<.0001
LBS100	1	-0.0194	0.0185	1.1035	0.2935
FOOTPRNT	1	-0.0466	0.0241	3.7569	0.0526
MINIVAN	1	0.5630	0.1575	12.7827	0.0003
CURTAIN	1	0.1313	0.1727	0.5780	0.4471
COMBO	1	-0.1543	0.1085	2.0239	0.1548
TORSO	1	-0.1590	0.1276	1.5543	0.2125
ABS	1	-0.1821	0.1860	0.9581	0.3277
ESC	1	-0.0546	0.1730	0.0996	0.7523
AWD	1	-0.0439	0.1198	0.1339	0.7144
DRVMALE	1	0.5340	0.1722	9.6109	0.0019
M14_30	1	0.0558	0.0187	8.9329	0.0028
M30_50	1	0.0125	0.00996	1.5693	0.2103
M50_70	1	0.0303	0.00954	10.0605	0.0015
M70_96	1	0.1230	0.0134	83.7119	<.0001
F14_30	1	0.0520	0.0209	6.1708	0.0130
F30_50	1	0.00619	0.0105	0.3491	0.5546
F50_70	1	0.0639	0.0112	32.5563	<.0001
F70_96	1	0.1162	0.0208	31.0815	<.0001
NITE	1	0.5267	0.0859	37.6319	<.0001
RURAL	1	1.6521	0.0762	470.2261	<.0001
SPDLIM55	1	2.2652	0.0770	865.7070	<.0001
HIFAT_ST	1	0.4821	0.0717	45.1985	<.0001
VEHAGE	1	0.0705	0.0241	8.5981	0.0034
BRANDNEW	1	0.2248	0.1207	3.4672	0.0626
CY2002	1	-0.1277	0.1735	0.5416	0.4618
CY2003	1	-0.2350	0.1560	2.2676	0.1321
CY2004	1	-0.1781	0.1363	1.7062	0.1915
CY2005	1	-0.1296	0.1203	1.1615	0.2811
CY2007	1	-0.1415	0.1085	1.7019	0.1920
CY2008	1	-0.2477	0.1150	4.6362	0.0313

The 1.22 trillion VMT of the CUVs and minivans were allocated among 371,009 induced-exposure records. They experienced 680 fatal collisions with heavy trucks, resulting in 828 fatalities (almost all in the CUVs or minivans). The VMT and the counts of fatal and induced-exposure crashes are much smaller than in the two preceding regressions. Fatality risk decreased by 1.94 percent for each additional 100 pounds and by 4.66 percent for each additional square foot of footprint in the CUVs and minivans; neither effect reaches statistical significance with the available data. In this type of crash, minivans have higher risk than CUVs after controlling for the other variables. The various types of side air bags, ABS, ESC, and AWD are generally beneficial. Drivers over age 70 are highly prone to colliding with heavy trucks. The crashes are especially prevalent in rural areas and on roads with speed limit 55+.

3.3 Mass and footprint coefficients for the 27 basic regressions

Table 3-1 presents the effects of <u>reducing</u> mass (while holding footprint constant) or reducing footprint (while holding mass constant) in 27 basic regressions: three vehicle classes x nine crash types. They estimate the average change in societal fatality risk for vehicles weighing n-100 pounds relative to vehicles weighing n pounds, keeping footprint and other factors constant and the change in risk for vehicles with footprint m-1 square feet relative to vehicles with footprint m square feet, keeping mass and other factors constant. Because the regressions for cars and truck-based LTVs each generate two coefficients for mass, one below and one above the median, the 27 regressions generate a total of 45 mass coefficients (but the footprint coefficient is the same for the lighter and heavier vehicles of the same class). These are <u>opposites</u> of the coefficients generated by the SAS® regressions, converted to percents – i.e., a negative number indicates that mass reduction is beneficial (or higher mass is associated with increased risk) and a positive number says mass reduction is harmful (or higher mass is associated with decreased risk). The Wald chi-square, as explained above, assumes the data are SRS and is a screening tool rather than an actual significance test: if it is less than 3.84, the coefficient is almost certainly not significant, after taking the actual sample design into account. If it exceeds 3.84, further analysis is needed to see if the coefficient is significant after accounting for the sampling design.

TABLE 3-1: EFFECTS OF MASS OR FOOTPRINT REDUCTION BY CASE-VEHICLE CLASS AND CRASH TYPE

CRASH TYPE	100-POUND MASS REDUCTION		1 SQ FT FOOTPRINT REDUCTION	
	FATALITY INCREASE (%)	WALD CHI-SQUARE	FATALITY INCREASE (%)	WALD CHI-SQUARE
CARS < 3,106 POUNDS				
1st-EVENT ROLLOVER	- 1.83	2.21	8.08	31.56
HIT FIXED OBJECT	- .46	.46	4.01	26.07
HIT PEDESTRIAN/BIKE/MOTORCYCLE	2.03	5.75	.91	.92
HIT HEAVY VEHICLE	2.26	3.77	2.97	4.86
HIT CAR/CUV/MINIVAN < 3082	.76	.57	.23	.04
HIT CAR/CUV/MINIVAN 3082+	.48	.25	.49	.20
HIT TRUCK-BASED LTV < 4150	1.17	.98	3.96	8.30
HIT TRUCK-BASED LTV 4150+	6.06	29.80	1.77	1.89
ALL OTHERS	1.95	9.96	1.14	2.64

	100-POUND MASS REDUCTION		1 SQ FT FOOTPRINT REDUCTION	
CRASH TYPE	FATALITY INCREASE (%)	WALD CHI-SQUARE	FATALITY INCREASE (%)	WALD CHI-SQUARE
CARS ≥ 3,106 POUNDS				
1st-EVENT ROLLOVER	-2.89	3.09	8.08	31.56
HIT FIXED OBJECT	-1.29	2.34	4.01	26.07
HIT PEDESTRIAN/BIKE/MOTORCYCLE	-.14	.02	.91	.92
HIT HEAVY VEHICLE	.39	.08	2.97	4.86
HIT CAR/CUV/MINIVAN < 3082	.26	.05	.23	.04
HIT CAR/CUV/MINIVAN 3082+	1.62	2.09	.49	.20
HIT TRUCK-BASED LTV < 4150	.53	.14	3.96	8.30
HIT TRUCK-BASED LTV 4150+	2.34	3.11	1.77	1.89
ALL OTHERS	1.16	2.51	1.14	2.64
PICKUPS & TRUCK-BASED SUVs < 4,594 POUNDS				
1st-EVENT ROLLOVER	.66	1.40	1.19	7.65
HIT FIXED OBJECT	-1.39	7.03	1.99	26.73
HIT PEDESTRIAN/BIKE/MOTORCYCLE	1.07	3.41	-1.24	8.60
HIT HEAVY VEHICLE	1.62	3.60	.75	1.41
HIT CAR/CUV/MINIVAN < 3082	-.09	.02	-.21	.24
HIT CAR/CUV/MINIVAN 3082+	-.71	1.30	.31	.46
HIT TRUCK-BASED LTV < 4150	-.63	.56	1.01	2.81
HIT TRUCK-BASED LTV 4150+	4.46	26.58	-1.69	6.88
ALL OTHERS	.73	2.86	-.44	1.93
PICKUPS & TRUCK-BASED SUVs ≥ 4,594 POUNDS				
1st-EVENT ROLLOVER	-1.28	7.51	1.19	7.65
HIT FIXED OBJECT	.76	2.74	1.99	26.73
HIT PEDESTRIAN/BIKE/MOTORCYCLE	-.05	.01	-1.24	8.60
HIT HEAVY VEHICLE	.32	.17	.75	1.41
HIT CAR/CUV/MINIVAN < 3082	-.91	3.99	-.21	.24
HIT CAR/CUV/MINIVAN 3082+	-1.37	8.05	.31	.46
HIT TRUCK-BASED LTV < 4150	-.96	1.99	1.01	2.81
HIT TRUCK-BASED LTV 4150+	.53	.53	-1.69	6.88
ALL OTHERS	-.11	.10	-.44	1.93
CUVs & MINIVANS				
1st-EVENT ROLLOVER	-7.02	15.31	11.59	22.66
HIT FIXED OBJECT	-3.61	7.47	7.67	18.19
HIT PEDESTRIAN/BIKE/MOTORCYCLE	-1.57	2.19	.37	.07
HIT HEAVY VEHICLE	1.94	1.10	4.66	3.76
HIT CAR/CUV/MINIVAN < 3082	-.09	.01	-.79	.19
HIT CAR/CUV/MINIVAN 3082+	1.68	1.54	-2.19	1.47
HIT TRUCK-BASED LTV < 4150	3.82	3.94	-4.05	2.48
HIT TRUCK-BASED LTV 4150+	-.93	.29	3.80	2.64
ALL OTHERS	-.40	.23	2.72	5.98

The principal findings in Table 3-1 are:

- In run-off-road crashes – first-event rollovers and impacts with fixed objects – mass reduction is usually beneficial (as evidenced by negative effects) and footprint reduction is always harmful (as evidenced by positive effects). Conversely, increased mass would tend to be harmful and added footprint protective in these crashes.
    - The four mass effects for passenger cars are not statistically significant, as evidenced by chi-square < 3.84.
    - In both subgroups of pickup trucks and truck-based SUVs, one mass effect is negative with chi-square > 3.84 (i.e., possibly significant after accounting for the sample design) while the other is positive and non-significant.
    - In CUVs and minivans, the estimated effects of mass reduction are strongly beneficial and of footprint reduction, exceedingly harmful: these may be possible symptoms of inaccurate estimates due to the near multicollinearity of mass and footprint.
    - Although footprint is protective in all vehicle classes, its estimated effect is stronger in cars, CUVs, and minivans than in pickup trucks and truck-based SUVs.
- In the other seven types of crashes, most of which take place on the road, mass reduction in the lighter vehicles usually creates societal harm; in the heavier vehicles, mass reduction creates societal benefits; and in the vehicles of medium weight, it has no consistent effect. Footprint reduction is consistently harmful in cars but has no consistent effect in pickup trucks, truck-based SUVs, CUVs, or minivans.
    - In the lighter cars (< 3,106 pounds), all seven mass effects are positive, and three of them have chi-square > 3.84.
    - In the heavier pickup trucks and SUVs (4,594 pounds and up), both mass effects for collisions with cars, CUVs, and minivans are negative with chi-square > 3.84.
    - In the lighter pickups and SUVs and the CUVs and minivans, seven mass effects are negative and seven are positive.
    - Footprint effects for pickups, SUVs, CUVs, and minivans are about half positive and half negative for these seven crash types.
- In collisions with pedestrians, bicyclists, and motorcyclists, the fatality rate increases as mass decreases in the lighter cars ($\chi^2 > 3.84$).
- In all classes of cars and LTVs, mass and footprint reductions are both harmful in collisions with heavy trucks.
- The strongest harmful effect for mass reduction, 6.06 percent fatality reduction per 100 pounds, may be found in cars < 3,106 pounds when they hit truck-based LTVs ≥ 4,150 pounds.
- The strongest beneficial effects for mass reduction, ranging from 0.91 to 1.37 percent fatality reduction per 100 pounds, may be found in pickups and SUVs ≥ 4,594 pounds when they hit lighter vehicles (but this range is well below the 6.06 effect in the lighter cars hitting the heavier LTVs).
- The results for CUVs and minivans, based on less data, vary more than the others. For example, in the last seven crash types, the estimated footprint effect can be positive, negative, or near zero.

Discussion: In general terms, these results are consistent with the hypotheses discussed in Section 1.6 and also consistent with the regressions of NHTSA's 2010 report, which are summarized in Table 3-2 for passenger cars (excluding 2-door muscle cars and 4-door police cars) and LTVs. Footprint reduction is harmful and mass reduction generally beneficial in rollovers and fixed-object collisions, consistent with handling and stability considerations. In the other types of crashes, mass reduction is harmful in the lighter cars and beneficial in the heavier LTVs, consistent with momentum considerations.

TABLE 3-2: 2010 REPORT[78] (MY 1991-1999 VEHICLES IN CY 1995-2000) - EFFECTS OF MASS OR FOOTPRINT REDUCTION BY CASE-VEHICLE CLASS AND CRASH TYPE

CRASH TYPE	100-POUND MASS REDUCTION		1 SQ FT FOOTPRINT REDUCTION	
	FATALITY INCREASE (%)	WALD CHI-SQUARE	FATALITY INCREASE (%)	WALD CHI-SQUARE
CARS < 2,950 POUNDS				
1st-EVENT ROLLOVER	-1.71	2.70	9.33	74.52
HIT FIXED OBJECT	-.40	.53	2.37	16.65
HIT PEDESTRIAN/BIKE/MOTORCYCLE	3.24	22.41	-.35	.26
HIT HEAVY VEHICLE	4.11	25.92	1.55	3.08
HIT CAR	1.10	4.58	1.03	3.47
HIT LTV	3.84	61.38	1.47	7.51
CARS ≥ 2,950 POUNDS				
1st-EVENT ROLLOVER	-1.18	1.15	9.33	74.52
HIT FIXED OBJECT	-1.32	5.88	2.37	16.65
HIT PEDESTRIAN/BIKE/MOTORCYCLE	-.48	.56	-.35	.26
HIT HEAVY VEHICLE	.97	1.52	1.55	3.08
HIT CAR	.94	3.58	1.03	3.47
HIT LTV	1.95	15.95	1.47	7.51
LTVs < 3,870 POUNDS				
1st-EVENT ROLLOVER	-4.61	63.76	7.10	361.48
HIT FIXED OBJECT	.08	.02	3.17	122.05
HIT PEDESTRIAN/BIKE/MOTORCYCLE	.51	.81	.57	3.20
HIT HEAVY VEHICLE	4.43	35.69	1.16	7.34
HIT CAR	-.17	.26	1.02	30.96
HIT LTV	3.00	25.97	.38	1.41
LTVs ≥ 3,870 POUNDS				
1st-EVENT ROLLOVER	-4.94	96.66	7.10	361.48
HIT FIXED OBJECT	-.55	1.55	3.17	122.05
HIT PEDESTRIAN/BIKE/MOTORCYCLE	-.48	1.05	.57	3.20
HIT HEAVY VEHICLE	-.67	1.05	1.16	7.34
HIT CAR	-1.78	44.84	1.02	30.96
HIT LTV	-1.92	16.51	.38	1.41

---

[78] Kahane (2010), Section 4.2 of Table 2-4 on p. 488 and Table 3-2 on p. 525.

There are some subtle changes, though, from NHTSA's 2010 results. The strong rollover-reducing effect of mass reduction for LTVs in the 2010 report now appears only for CUVs and minivans.[79] Footprint reduction is no longer consistently harmful for LTVs in crashes other than rollovers and impacts with fixed objects. In several crash types (pedestrians, heavy trucks, car-to-car), the harmful effects of mass reduction in the lighter cars have diminished to some extent. In the heavier LTVs, the benefits of mass reduction have likewise diminished in their crashes with cars. In other words, the results appear to be directionally similar to the 2010 report but somewhat weaker in magnitude. These findings will be discussed further in Section 3.5, after estimation of the overall effect of mass reduction.

3.4 The effect of mass within deciles of footprint

The relatively high correlation and VIF of mass and footprint raised questions about near multicollinearity. A few of the regressions in Table 3-1 displayed possible symptoms of multicollinearity, namely a strong positive coefficient for mass and a strong negative coefficient for footprint – or vice versa. Specifically, the CUV/minivan regressions for rollovers and fixed-object crashes displayed strong positive coefficients for mass (7.04, 3.61) and strong negative coefficients for footprint (-11.59, -7.67), whereas the CUV/minivan regressions for collisions with LTVs < 4,150 pounds was strongly negative for mass (-3.82) and positive for footprint (4.05).[80] Directionally similar, but with weaker estimated effects for mass, are the car regression for rollovers (mass coefficients 1.83 and 2.89, footprint coefficient -8.08) and, with opposite signs, the CUV/minivan regression for collisions with cars ≥ 3,082 pounds (mass coefficient -1.68, footprint coefficient 2.19).

Farmer and Green in their peer reviews of the 2010 report recommended more attention to the issue. However, as discussed in Section 4.5, attempts to index mass to footprint, such as "excess footprint given mass" or vice-versa do not meaningfully change the results of the logistic regressions. Another way to avoid having mass and footprint in the same regression is to split up the database into deciles of footprint. Within a decile, all the cars have similar footprint, typically within a square foot. Within each decile, logistic regressions can be performed with curb weight and all the other independent variables except footprint – because footprint is virtually constant, not a variable, within a decile. The 2010 report presented numerous decile analyses, but most of them considered only the correlation of mass and the simple fatality rate.[81] They did not take into account the interaction of mass with driver age/gender or other factors, as Farmer pointed out in his peer review. But one decile analysis for passenger cars in the 2010 report[82] did control for these other factors by running the logistic regressions, without footprint, in each decile – and that is the method that will be used here.

The logistic regressions within each decile of footprint will use the same variables as in the preceding section, except footprint. Also, in the analyses of cars and truck-based LTVs, the two-

---

[79] Even in the 2010 report, the effect changed directions when the LTV regressions were limited to pickup trucks, raising questions about the reliability of the estimate; see Kahane (2010), pp. 531-532.
[80] Table 3-1 shows the opposite sign for each regression coefficient, because it estimates the effect of reducing mass or footprint.
[81] Kahane (2010), pp. 506-512 and 527-531.
[82] *Ibid.*, Table 2-13 on p. 511.

piece linear curb-weight variables are replaced by the simple, linear variable LBS100, as in the basic analyses of CUVs and minivans. The objective, for each vehicle class and type of crash, is to find out in how many deciles the regression estimates a negative coefficient for LBS100 (suggesting mass reduction is harmful), and how many positive (suggesting mass reduction is beneficial). (The chance of the coefficient being exactly zero is infinitesimal.) The analysis just counts the numbers of negative and positive coefficients and does not take into account whether the coefficients are statistically significant. NHTSA considers the decile analysis a relatively blunt, essentially non-parametric tool (i.e., simply counting how many negative coefficients) for just confirming the directional accuracy of the basic regressions. Although a more extensive analysis is theoretically feasible – e.g., using the mass coefficients for each decile to compute an overall effect – it is not clear what advantage it would have over the basic regressions and, because footprint still varies to some extent within each decile, it would not be an analysis that fully controls for footprint, either.

Table 3-3 enumerates the ten deciles of footprint and specifies the range of footprint and curb weight in each decile: for passenger cars (excluding muscle cars, police cars, and cars with AWD), for pickup trucks and truck-based SUVs, and for CUVs and minivans. Within the middle deciles, footprints for cars are in a range of about one square foot, while curb weights vary by 1,000 pounds; footprints for pickup trucks and truck-based SUVs are in a range of about 3 or 4 square feet while curb weights vary by over 2,000 pounds.

TABLE 3-3: TEN DECILES OF FOOTPRINT FOR MY 2000-2007 VEHICLE GROUPS

Footprint Deciles	Range of Footprint (Square Feet)	Range of Curb Weight (Pounds)
PASSENGER CARS		
1st	34.8 to 40.5	1799 to 3283
2nd	40.6 to 41.4	2359 to 3167
3rd	41.5 to 42.1	2406 to 3424
4th	42.3 to 43.5	2559 to 3883
5th	43.6 to 44.0	2727 to 3705
6th	44.1 to 45.0	2916 to 3838
7th	45.1 to 46.3	2954 to 4019
8th	46.5 to 46.9	3109 to 4020
9th	47.0 to 48.2	3346 to 4277
10th	48.3 to 55.5	3486 to 4562

PICKUP TRUCKS AND TRUCK-BASED SUVs

1st	34.6 to 43.2	2580 to 4704
2nd	43.3 to 45.2	2960 to 4527
3rd	45.3 to 48.2	3198 to 5566
4th	48.3 to 50.4	2833 to 5758
5th	50.6 to 54.0	3159 to 5573
6th	54.1 to 55.8	3835 to 6070
7th	56.1 to 61.0	3558 to 6642
8th	61.6 to 65.2	3865 to 7346
9th	65.3 to 66.8	4266 to 6393
10th	67.1 to 81.0	4266 to 7520

CUVs AND MINIVANS

1st	38.0 to 42.5	2712 to 3699
2nd	42.8 to 43.6	3089 to 3831
3rd	43.7 to 45.1	3086 to 3929
4th	45.2 to 47.4	3205 to 4275
5th	47.8 to 49.4	3660 to 4828
6th	49.5 to 50.2	3559 to 4687
7th	50.4 to 52.5	3838 to 5335
8th	= 52.6	3824 to 4651
9th	53.3 to 53.5	3719 to 4677
10th	53.8 to 58.0	4071 to 5249

Table 3-4 shows in how many of the ten deciles the regression coefficient for curb weight (LBS100) is negative – i.e., implies that mass reduction is harmful. (In the remainder of the ten deciles, the coefficient is positive and implies that mass reduction is beneficial.) Table 3-4 just counts how many of the 10 individual coefficients are negative; they are not necessarily statistically significant. Examples of inconsistency between Table 3-4 and the basic regression results (which include the footprint variable) in Table 3-1 – possibly indicating that the basic regression inaccurately allocates the relative contributions of mass and footprint due to their near multicollinearity – could include:

- Mass reduction is strongly <u>beneficial</u> in the basic regression – as evidenced by a substantial negative "fatality increase (%)" per 100-pound mass reduction in Table 3-1, with an accompanying Wald chi-square well above 3.84[83] – but <u>harmful</u> in most of the deciles – as evidenced by, say, a 7 or more in Table 3-4
- Mass reduction is strongly <u>harmful</u> in the basic regression – as evidenced by a substantial positive "fatality increase (%)" per 100-pound mass reduction in Table 3-1, with an

---

[83] For passenger cars and truck-based LTVs, where there are actually two mass coefficients in Table 3-1 – e.g., one for cars < 3,106 and one for cars ≥ 3,106 – these two coefficients should be in the same direction and at least one of them have Wald chi-square well above 3.84.

accompanying Wald chi-square well above 3.84 – but <u>beneficial</u> in most of the deciles – as evidenced by a 3 or less in Table 3-4
- A near-zero or non-significant effect in the basic regressions[84] but the decile analyses lean strongly either way, as evidenced by a 0, 1, 9, or 10 in Table 3-4.

TABLE 3-4: NUMBER OF FOOTPRINT DECILES IN WHICH MASS REDUCTION IS HARMFUL, BY CASE VEHICLE CLASS AND CRASH TYPE
(Number of deciles where mass coefficient is negative, not necessarily statistically significant)

Crash Type	Passenger Cars	Pickups & Truck-Based SUVs	CUVs & Minivans
First-event rollovers	5	6	4
Hit fixed object	4	5	5
Hit pedestrians/bikes/motorcycles	7	5	3
Hit heavy truck or bus	7	7	6
Hit car, CUV or minivan < 3,082 lbs.	5	3	5
Hit car, CUV or minivan ≥ 3,082 lbs.	5	3	8
Hit truck-based LTV < 4,150 lbs.	5	6	7
Hit truck-based LTV ≥ 4,150 lbs.	7	9	5
All other crash involvements	6	5	6

The basic regressions and decile analyses are generally consistent for passenger cars, except perhaps in rollovers. The basic regressions said mass reduction was beneficial (but not statistically significant) but mass reduction is harmful (has a negative coefficient) in 5 of the 10 deciles; that is not a glaring inconsistency, given that the basic regression results were not significant, but neither does the decile analysis support a conclusion that mass reduction is beneficial. The basic regressions said mass reduction was beneficial (but not statistically significant) in collisions with fixed objects and, here, mass reduction is harmful in only 4 deciles. Mass reduction was generally harmful in pedestrian, heavy truck, LTV ≥ 4,150, and "all other" crash involvements in Table 3-1 and, here too, mass reduction is harmful in 7, 7, 7, and 6 of the 10 respective deciles. But, specifically, mass reduction was especially harmful in Table 3-1 for the cars < 3,106 pounds in pedestrian, heavy truck, LTV ≥ 4,150 collisions – and here, mass reduction was harmful in 4 of the 5 lowest footprint deciles in pedestrian collisions and in all 5 of the lowest footprint deciles in collisions with heavy trucks and LTVs ≥ 4,150 pounds. Mass had little societal effect in the collisions with other cars in Table 3-1 and the decile analysis is likewise 50-50 in both cases.

For pickup trucks and truck-based SUVs, the basic regression said mass reduction was societally beneficial in collisions with cars – and the decile analysis likewise says mass reduction is harmful in only 3 deciles in each type of collisions with cars. Furthermore, in both cases, mass reduction is beneficial in 4 of the 5 highest footprint deciles. The basic regression showed a strong effect for UNDRWT00 (harm for mass reduction in the lighter LTVs) in collisions with

---

[84] Or effects in opposite directions for the lighter and heavier vehicles of the same type.

heavier LTVs and this is echoed by harm in 9 deciles, including all 5 of the lowest footprint deciles. In collisions with heavy trucks, mass reduction in the LTVs had a slightly harmful effect in Table 3-1, and that is consistent with the decile analysis (harmful effect in 7 deciles). In the remaining types of crashes, mass reduction had an inconsistent or marginal effect, and that is also consistent with the decile analyses (harmful effect in 5 or 6 deciles).

The analyses for CUVs and minivans generated harmful effects (negative coefficients) for mass reduction in 4 deciles for rollovers and in 5 deciles for collisions with fixed objects. Those results do not fully support the strong benefits found for mass reduction in the basic regressions, but they do not unequivocally contradict them, either. On the other hand, in the two types of crashes where the basic regressions showed harm for mass reduction and a benefit for footprint reduction (collisions with cars ≥ 3,082 and LTVs < 4,150), the decile analyses showed harm for mass reduction in 8 and 7 deciles, respectively. Thus, in four basic regressions displaying opposite signs for mass and footprint, the decile analyses do not provide clear evidence that these effects are – or are not – likely consequences of multicollinearity. In the five other types of crashes, the decile analyses are consistent with the basic regressions.

Table 3-4 includes 270 individual regression coefficients for deciles, 90 each for cars, truck-based LTVs, and CUVs/minivans. Because coefficients are not statistically independent across crash types (as different crash types use the same induced-exposure data), the tallies of positive and negative coefficients cannot be statistically tested as if they were 270 independent observations. Overall, 51 of the 90 coefficients for cars are in the direction of mass reduction being harmful, 49 of the 90 coefficients for truck-based LTVs, and 49 of the 90 coefficients for CUVs/minivans. In other words, the deciles split close to 50-50 for all three groups of vehicles, but leaning somewhat in the direction of mass reduction being harmful in passenger cars. The only cases where the decile analysis consistently shows mass reduction being beneficial is for truck-based LTVs when they hit cars, CUVs and minivans. The decile analysis does not indicate a clear benefit for mass reduction in rollovers and fixed-object crashes, but it does not rule it out, either.

In the big picture, the decile analyses do not raise any serious doubts about the results of the basic regressions; none of the 27 analyses shows a strong inconsistency between the basic regression coefficients and the decile coefficients, based on the three criteria listed before Table 3-4. They suggest that the basic regressions, <u>on average</u>, get the relative effects of mass and footprint about right. There does not appear to be a systematic bias against mass and in favor of footprint, or vice-versa. The decile analyses perhaps raise doubts about individual regressions, specifically CUV rollovers and CUV-fixed-object, where the benefit attributed to mass reduction is quite large in the basic regressions, but these possible symptoms of multicollinearity appear to be offset and essentially cancelled out by symptoms of the opposite direction, namely a high benefit for footprint reduction in the CUV-heavy car and CUV-light LTV regressions. Separating the CUVs and minivans from the other LTVs may also have helped stabilize the results. NHTSA has no other explanation of why multicollinearity became less of a problem, except this: when there are only a few (2-4) regressions in each report that seem to display symptoms of multicollinearity, namely exceptionally strong coefficients in opposite directions for mass and footprint, it could readily happen by chance that all of them give the positive coefficient to the same variable in one report (curb weight in 2010) but split 50-50 in this report.

The decile analyses encourage accepting the sundry results of the basic regression analyses as they are. Unlike the 2010 report (see pp. 514-520 and 532-535), it does not appear useful to propose additional "upper-estimate" and "lower-estimate" scenarios that set aside some of the regression coefficients and replace them with estimates derived from other sources. Instead, the range of estimates in this report will be the confidence bounds on the actual regression results.

3.5 Overall effect of mass reduction by vehicle class and its confidence bounds

The John A. Volpe National Transportation Systems Center of the United States Department of Transportation has a computer model, the CAFE Compliance and Effects Modeling System (usually referred to as the "Volpe model"), that works out the impacts of CAFE standards, including safety effects, over the lifetimes of future vehicles. The Volpe model requires five basic numbers in order to predict the safety effects, if any, of foreseeable mass reductions in MY 2017+ vehicles. The five numbers are the overall percentage increases or decreases, per 100-pound mass reduction while holding footprint constant, in crash fatalities (including the occupants of other vehicles and non-occupants) involving case vehicles that are:

- Passenger cars weighing less than 3,106 pounds
- Passenger cars weighing 3,106 pounds or more
- Truck-based LTVs weighing less than 4,594 pounds
- Truck-based LTVs weighing 4,594 pounds or more
- CUVs and minivans

Table 3-5 computes these five percentages: point estimates and also upper and lower 95% sampling-error confidence bounds, which will serve as ranges for the estimates. These confidence bounds take the sample design into account and are not based on the Wald chi-square statistics in Table 3-1 (which assume SRS). For example, in passenger cars < 3,106 pounds, the point estimate is that crash fatalities would increase by 1.56 percent per 100-pound mass reduction, confidence bounds ranging from a 0.39 to a 2.73 percent increase; but in truck-based LTVs ≥ 4,594 pounds, societal fatalities would decrease by a non-significant 0.34 percent per 100-pound mass reduction, confidence bounds ranging from a 0.97 percent decrease to 0.30 percent increase.

TABLE 3-5: ESTIMATED EFFECTS OF 100-POUND MASS REDUCTION WHILE HOLDING FOOTPRINT CONSTANT
BY VEHICLE CLASS AND CRASH TYPE

CRASH TYPE	FATALITIES AFTER ESC	POINT ESTIMATE		95% CONFIDENCE BOUNDS (%)	
		N	%	LOWER	UPPER
CARS < 3,106 POUNDS					
1st-EVENT ROLLOVER	207	-4	-1.83	-4.32	.66
HIT FIXED OBJECT	813	-4	-.46	-2.19	1.27
HIT PEDESTRIAN/BIKE/MOTORCYCLE	871	18	2.03	.15	3.91
HIT HEAVY VEHICLE	471	11	2.26	-1.16	5.68
HIT CAR-CUV-MINIVAN < 3082	478	4	.76	-1.55	3.06
HIT CAR-CUV-MINIVAN 3082+	674	3	.48	-2.33	3.29
HIT TRUCK-BASED LTV < 4150	351	4	1.17	-1.93	4.27
HIT TRUCK-BASED LTV 4150+	505	31	6.06	3.26	8.86
ALL OTHERS	1,530	30	1.95	-.30	4.20
OVERALL	5,901	92	1.56	.39	2.73
CARS ≥ 3,106 POUNDS					
1st-EVENT ROLLOVER	247	-7	-2.89	-6.24	.47
HIT FIXED OBJECT	1,263	-16	-1.29	-3.08	.50
HIT PEDESTRIAN/BIKE/MOTORCYCLE	1,388	-2	-.14	-1.75	1.47
HIT HEAVY VEHICLE	687	3	.39	-3.55	4.33
HIT CAR-CUV-MINIVAN < 3082	921	2	.26	-2.20	2.72
HIT CAR-CUV-MINIVAN 3082+	1,172	19	1.62	-1.50	4.74
HIT TRUCK-BASED LTV < 4150	543	3	.53	-2.53	3.59
HIT TRUCK-BASED LTV 4150+	727	17	2.34	-.80	5.48
ALL OTHERS	2,552	30	1.16	-1.37	3.70
OVERALL	9,499	48	.51	-.59	1.60
PICKUPS & TRUCK-BASED SUVs < 4,594 POUNDS					
1st-EVENT ROLLOVER	162	1	.66	-1.23	2.54
HIT FIXED OBJECT	451	-6	-1.39	-2.97	.20
HIT PEDESTRIAN/BIKE/MOTORCYCLE	676	7	1.07	-.58	2.72
HIT HEAVY VEHICLE	287	5	1.62	-.47	3.71
HIT CAR-CUV-MINIVAN < 3082	530	-0	-.09	-1.73	1.55
HIT CAR-CUV-MINIVAN 3082+	485	-3	-.71	-2.25	.84
HIT TRUCK-BASED LTV < 4150	252	-2	-.63	-2.93	1.68
HIT TRUCK-BASED LTV 4150+	289	13	4.46	2.14	6.78
ALL OTHERS	1,126	8	.73	-.70	2.17
OVERALL	4,258	22	.52	-.45	1.48

		FATALITY INCREASE PER 100-POUND MASS REDUCTION			
CRASH TYPE	FATALITIES AFTER ESC	POINT ESTIMATE N	%	95% CONFIDENCE BOUNDS (%) LOWER	UPPER
PICKUPS & TRUCK-BASED SUVs ≥ 4,594 POUNDS					
1st-EVENT ROLLOVER	345	-4	-1.28	-2.95	.39
HIT FIXED OBJECT	802	6	.76	-.27	1.80
HIT PEDESTRIAN/BIKE/MOTORCYCLE	1,516	-1	-.06	-1.10	.99
HIT HEAVY VEHICLE	492	2	.32	-1.38	2.02
HIT CAR-CUV-MINIVAN < 3082	1,262	-12	-.91	-2.25	.42
HIT CAR-CUV-MINIVAN 3082+	1,155	-16	-1.37	-2.88	.14
HIT TRUCK-BASED LTV < 4150	578	-6	-.96	-2.43	.51
HIT TRUCK-BASED LTV 4150+	490	3	.53	-2.48	3.55
ALL OTHERS	2,262	-3	-.11	-1.05	.82
OVERALL	8,902	-30	-.34	-.97	.30
CUVs & MINIVANS					
1st-EVENT ROLLOVER	100	-7	-7.02	-13.00	-1.03
HIT FIXED OBJECT	373	-13	-3.61	-6.34	-.89
HIT PEDESTRIAN/BIKE/MOTORCYCLE	812	-13	-1.57	-3.41	.27
HIT HEAVY VEHICLE	297	6	1.94	-3.91	7.79
HIT CAR-CUV-MINIVAN < 3082	503	-0	-.09	-2.87	2.69
HIT CAR-CUV-MINIVAN 3082+	569	10	1.68	-2.82	6.18
HIT TRUCK-BASED LTV < 4150	244	9	3.82	-1.61	9.24
HIT TRUCK-BASED LTV 4150+	294	-3	-.93	-6.58	4.73
ALL OTHERS	1,380	-5	-.40	-3.82	3.03
OVERALL	4,571	-17	-.37	-1.55	.81

The principal findings in Table 3-5 are that crash fatalities would increase by 1.56 percent per 100-pound mass reduction in passenger cars < 3,106 pounds (confidence bounds from 0.39% to 2.73%). In the other four vehicle classes, two point estimates are increases (0.51% in cars ≥ 3,106 pounds and 0.52% in truck-based LTVs < 4,594 pounds), two are fatality reductions (0.34% in truck-based LTVs ≥ 4,594 pounds and 0.37% in CUVs and minivans), but none are statistically significant, as evidenced by confidence bounds ranging from a reduction to an increase. Combined effects of simultaneously reducing mass in all five vehicle classes are estimated in the next section.

Here is how the statistics in Table 3-5 are derived: first, the point estimates. The point estimates expressed as percentages (the middle column of numbers) for the individual crash types are copied from the left column of numbers in Table 3-1 and are based on the actual regression results.

In order to obtain an overall effect across crash types, it is necessary to gauge the relative incidence of each type of crash: the "baseline" fatalities. As in NHTSA's 2003 and 2010 reports, the baseline is derived from a subset of the more recent fatalities in the FARS analysis database, namely the last four MY in the last five CY (MY 2004-2007 vehicles in CY 2004-2008 crashes).[85] The choice of the last four MY and last five CY has no special meaning but just represents one possible trade-off between the two conflicting goals of using the latest-possible vehicles and having enough cases in each cell to get a precise distribution; this choice retains approximately one-third of the original FARS cases. Furthermore, because all new vehicles already (as of April 2012) are equipped with ESC, the original baseline fatalities are adjusted downward to what they would have been if all vehicles had been equipped with ESC (as they will be in MY 2017-2025). NHTSA's most recent statistical evaluation estimates[86] that ESC reduces fatal first-event rollovers by 56 percent in cars and 74 percent in LTVs (including CUVs and minivans); fixed-object impacts by 47 percent in cars and 45 percent in LTVs; and other non-pedestrian crashes by 8 percent in both cars and LTVs. For example, if the database has 200 records of cars in fatal first-event rollovers and 100 of these cars were ESC-equipped, 100 not equipped, the baseline fatalities would be adjusted downward from 200 to 144 to reflect that the 100 fatalities in the non-equipped cars would have dropped by 56 percent, to 44, if they had been ESC-equipped (whereas the 100 fatalities in the already ESC-equipped cars would have stayed the same). Baseline fatalities have not been adjusted for other upcoming technologies, specifically curtain air bags, because they will not radically change the distribution of fatalities by crash type, only reduce the overall absolute number; they will not substantially change the relative weights of the nine crash types.

The baseline fatalities after ESC are the left column of numbers in Table 3-5. These are not annual fatality counts but are based simply on the actual counts of fatal-crash involvements for MY 2004-2007 vehicles in CY 2004-2008, for the purpose of averaging the effects of mass reduction across crash types. They are vehicle-based societal counts. For example, the 505 baseline fatalities for cars < 3,106 pounds in collisions with LTVs ≥ 4150 pounds indicates that cars were involved in a number of fatal collisions that resulted in a total of 505 fatalities in the cars or in the partner LTVs. If some of those partner LTVs were MY 2004-2007, the same crash would also appear in one of the baseline counts for LTVs hitting cars/CUVs/minivans. The double-counting is not an issue here (where separate effects are estimated for each class of vehicles) but needs to be addressed in Section 3.6 (where effects will be combined across vehicle classes). Section 3.6 will also index the fatality counts to annual totals to allow estimation of the effects of mass reduction on annual fatalities.

For each type of crash, the regression coefficient is applied to the baseline fatalities to estimate the numerical increase or decrease in fatalities for that crash type (the second column of numbers in Table 3-5). Within each vehicle class, the sum of the nine numerical increases or decreases divided by the sum of the baseline fatalities yields the overall percentage effect of a 100-pound mass reduction for that class of vehicles. For example, in cars < 3,106 pounds, the increase of 92 fatalities is 1.56 percent of the baseline, 5,901.

Confidence bounds: Two sources of sampling error are considered:

---

[85] Kahane (2003), p. 104; also, the vehicles not included in the regressions, namely 2-door muscle cars, 4-door police cars, and full-size vans are included in tabulating baseline fatalities.

[86] Sivinski R. (2011). *Update of NHTSA's 2007 Evaluation of the Effectiveness of Light Vehicle Electronic Stability Control (ESC) in Crash Prevention*, NHTSA Technical Report No. DOT HS 811 486. Washington, DC: National Highway Traffic Safety Administration. http://www-nrd.nhtsa.dot.gov/Pubs/811486.pdf.

- The relatively small numbers (hundreds or thousands, not hundreds-of-thousands or millions) of fatal-crash cases included in the regression analyses. Of course, FARS is technically a census, not a sample, but NHTSA analyses usually treat FARS data as if it were a sample and apply customary statistical tests such as chi-square. The crashes that actually occur in the course of a year are construed as a sample of the crashes that would have occurred if more-or-less the same national crash environment of that year had been repeated over and over, each year resulting in a somewhat different number and distribution of fatal crashes.
- The fact that induced-exposure data from only 13 States, rather than from all 50 States plus DC are used to allocate the registration years and VMT by the various control variables. With other States, the allocation might have been somewhat different. Technically, these 13 States were not selected by simple random sampling but are the States whose files are available to NHTSA and include the VIN and the other necessary control variables. But to the extent that the availability of these States' files rather than other States' does not appear to have been influenced by any criterion directly relevant to this study, these 13 States may be considered a quasi-random sample, at least for the purpose of assessing sampling error.

The first source of error ("FARS-based sampling error") far exceeds the "State-based sampling error" in the estimates of the individual regression coefficients. But as the results are averaged across crash types (in Table 3-5), the State-based errors, which have high covariance across crash types (because all regressions use the same exposure data), come closer to the FARS-based error, which has little covariance across crash types (because each regression uses a different set of FARS cases).

Both sources of error are estimated by a jackknife technique, because of the complexity of the estimator (a logistic regression coefficient) and the need for ample data to drive the regression. For the FARS-based error, the FARS cases are subdivided into 10 systematic random subsamples of equal size, based on the last digit of the case number, ST_CASE – i.e., at the accident, not the vehicle or person level; as Paul Green pointed out, the fatality cases are essentially "clustered" at the accident level. Ten regressions are performed, each using the 9/10 of the FARS data that remain after one of the subsamples is removed and using all the induced-exposure data. (The subsample is then replaced before the next subsample is removed.) The 10 regressions yield 10 estimates of the regression coefficients – specifically those for the mass variables (UNDRWT00, OVERWT00, or LBS100, depending on the vehicle class) – each of which is slightly different from the original coefficient based on the full FARS data. If, for example, the original coefficient is x and the coefficient is $x + h$ when all FARS cases are used except those with ST_CASE ending in zero, a "pseudo-estimate" $x - 9h$ is generated for the subsample including only the FARS cases with ST_CASE ending in zero (because if a regression could have been run using only these cases, it would have had to produce a coefficient $x - 9h$ in order for it and the $x + h$ generated from the other 9/10 of the data to average out to x). The standard error of these 10 pseudo-estimates serves as the FARS-based component of the standard deviation of the original coefficient and it can be treated as a t-distribution with 9 degrees of freedom (df). This FARS-based error is typically a little bit more than the error implied by the Wald chi-square statistics in Table 3-1; the slight increase may be due to FARS clustering at the accident level (more than one person may be a fatality in the same crash) whereas the Wald chi-square treats each fatally injured person as a separate, independent case.

For the State-based error, the induced-exposure data is construed as a cluster sample, each State plus DC constituting a primary sampling unit (PSU); 13 of the 51 available PSUs are in the sample. Unlike the FARS data, the PSUs cannot be partitioned into subsamples of equal size, because the States vary considerably in size (i.e., cumulative VMT). But if the three least populous States – Kansas, Nebraska, and Wyoming – are combined, there are 11 subsamples, each containing at least 3.7 percent of the cumulative VMT. Eleven regressions are performed, each using all of the FARS data, but using only the induced-exposure data from the remaining States, after the State(s) in that subsample are removed, to allocate the VMT by driver age, gender, and the other control variables, as explained in Section 2.6. In each of the regressions, the VMT still add up to the national totals; they are just allocated differently. The subsample is then replaced before the next subsample is removed. The 11 regressions yield 11 estimates of the regression coefficients for UNDRWT00, OVERWT00, and/or LBS100, each slightly different from the original coefficients based on using the induced-exposure data from all 13 States to allocate the VMT. If, for example, the original coefficient is x and the coefficient is $x + h$ when all States except Alabama are used to allocate the VMT and if the Alabama records were allocated a share w of the nation's VMT in the original file, a "pseudo-estimate" $x - h(w/(1-w))$ is generated for the subsample including only the Alabama cases (because if a regression could have been run using only Alabama induced-exposure cases to allocate the VMT, it would have had to produce a coefficient $x - h(w/(1-w))$ in order for it and the $x + h$ coefficient generated by using all the States except Alabama, which account for a 1-w share of the data, to have the share-weighted average be x). The share-weighted standard error of these 11 pseudo-estimates is multiplied by a finite population correction (FPC) of $.8718 = \sqrt{(51-13)/(51-1)}$, because the sample included 13 of the 51 available PSUs. This serves as the State-based component of the standard deviation of the original coefficient and it can be treated as a t-distribution with 10 degrees of freedom (df).

The standard deviation of the original coefficient is the root-sum-of-squares of the FARS- and State-based components. The two-sided 95% confidence bounds for the regression coefficients for the individual crash types are the point estimate of the coefficient (as derived from the basic regression analysis) ±2.262 standard deviations, where 2.262 is the 97.5th percentile of a t-distribution with 9 df (the lesser df of the FARS- and State-based components – i.e., the wider confidence bound). These are the confidence bounds shown for the individual crash types in Table 3-5. The t-test with 9 df can also be applied to test if the point estimate is statistically significant.

The same sets of pseudo-estimates for the individual regression coefficients can be used to compute confidence bounds for any linear combination of point estimates for these coefficients. For example, the point estimate of the overall effect of a 100-pound mass reduction in cars < 3,106 pounds, a 1.56 percent fatality increase according to Table 3-5, is the weighted average of the nine coefficients above it in the table, the "annual fatalities after ESC," two columns to the left, being the weighting factor. The overall effect is recomputed using, for each of the nine crash types, the pseudo-estimate coefficient for the FARS cases with ST_CASE ending in 0 substituted for the point-estimate coefficient (but the same weighting factor); again recomputed using the pseudo-estimates for the FARS cases with ST_CASE ending in 1; and so on. The standard error of the 10 resulting pseudo-estimates of the overall effect serves as the FARS-based component of its sampling error. The overall effect is likewise recomputed 11 times using the pseudo-estimate coefficients for the various State files to estimate the State-based component of its sampling error. The confidence bounds for the overall effects of 100-pound mass reduction in the five classes of vehicles, the "Volpe model coefficients," are also shown in Table 3-5. The overall 1.56 percent fatality increase per 100-pound mass reduction in the cars < 3,106 is statistically significant ($t = 3.02$, $df = 9$). The Volpe coefficients for the other four vehicle groups, .51, .52, -.34, and -.37 are not statistically significant ($t = 1.04$, 1.21, -1.20, and $-.71$, respectively). Because the confidence bounds and the significance test are based on similar t-statistics, the confidence bounds will exclude zero when and if the t-test with 9 df is significant ($t > 2.262$).

In his peer review of Wenzel's 2011 report, Greene questions the significance of the effect in cars < 3,106 pounds, given that five vehicle groups are tested and only one result was significant at the .05 level. "When testing a hypothesis on, for example, 5 vehicle classes simultaneously, a result for one equation that might be statistically significant on its own may not be statistically significant as one of five related tests."[87] Greene's discussion would apply if the five subgroups were selected randomly, or even by a non-random criterion if there were no reason to believe that mass reduction might have different effects in the various subgroups. In this case, though, there are distinct hypotheses on the potential societal effects of mass reduction, depending on the mass of the vehicle (see Section 1.6). The hypotheses are that mass reduction in the lightest vehicles would likely increase societal risk, while mass reduction in the heaviest vehicles could have benefits – and the results are consistent with the hypotheses. In that sense, the analyses of the five vehicle classes can be seen as five tests of separate hypotheses rather than five simultaneous tests of the same hypothesis.

Discussion of results: The results with the new database are directionally the same as in NHTSA's 2010 report. In the lighter cars, mass reduction is associated with societal harm, in the heavier LTVs, with societal benefits, and in the other vehicles with little effect in either direction. The results continue to be consistent with the idea, based on momentum considerations, that it does more harm than good to make the lightest vehicles even lighter but that it does more good than harm to make the heaviest vehicles lighter.

What may have changed is that the fatality increase in the lighter cars and the reduction in the heaviest LTVs are of smaller magnitude than the previous database. This conclusion is not definite because of the relatively wide confidence bounds of the estimates. Only the overall effect in the lighter cars remains statistically significant. Only five of the 45 coefficients for individual crash types in Table 3-5 are statistically significant, as evidenced by upper and lower bounds with the same sign: the fatality increase for mass reduction when cars < 3,106 pounds in collisions with pedestrians, the fatality increases for cars < 3,106 pounds and LTVs < 4,594 pounds in collisions with LTVs ≥ 4,150 pounds, and the benefits of mass reduction in CUV/minivan rollovers and fixed-object impacts. Only 11 of the 45 coefficients had Wald chi-

---

[87] Menard, B., ed. (2012). *Peer Review of LBNL Statistical Analysis of the Effect of Vehicle Mass & Footprint Reduction on Safety (LBNL Phase 1 and 2 Reports)* (To appear in Docket No. NHTSA-2010-0152). Charlottesville, VA: Systems Research and Application Corp.

squares > 3.84 and only three had $\chi^2 > 10$ (see Table 3-1). By contrast, 12 of the 24 coefficients in the 2010 analysis had $\chi^2 > 3.84$ (see Table 3-2 of this report) and 10 had $\chi^2 > 10$.

Four factors may be suggested for why the trends may have become weaker. The first three pertain to actual changes in vehicle design (MY 2000-2007 versus MY 1991-1999) or driving patterns (CY 2002-2008 versus CY 1995-2000); the last relates to the statistical method:

- Many vehicles with poor safety performance, as evidenced, for example, in the IIHS offset tests, had been discontinued by 2000 or at the latest during 2000-2007. These poor performers were often light, small cars, as discussed in Section 1.4.
- Blocker beams and other voluntary compatibility improvements in LTVs as well as compatibility-related self-protection improvements to cars have made the heavier LTVs less aggressive in collisions with lighter vehicles. The high societal fatality rates of these vehicles in collisions with lighter vehicles in the earlier database may have reflected the extra aggressiveness of the LTVs and the extra vulnerability of the lighter partners in addition to the effects of mass *per se*.
- The tendency of lighter and/or small vehicles to be driven poorly, while still there, is not as strong as it used to be, as evidenced by the analysis of culpability in fatal crashes discussed in Section 1.7. Perhaps some of the negative perceptions of small, light cars have diminished as the product has improved, in turn diminishing the "self-fulfilling prophecy" of better drivers avoiding these vehicles.
- Analysis of CUVs and minivans as a class of vehicles separate from truck-based LTVs may have avoided possible inaccuracies in the 2010 regression results for truck-based LTVs, CUVs, and minivans combined: specifically, the earlier report's findings of strong benefits of mass reduction in rollovers and in collisions with cars. The strong effect in rollovers is now confined to the CUVs and minivans and the effect in collisions with cars is now weaker in each of the separate analyses. As Paul Green noted in his peer review, including many disparate LTVs, each with its own pattern of crash types and of relationships between mass and footprint, might generate coefficients for mass and footprint that reflect the vehicle mix rather than the underlying relationships, within each individual type of LTV, of these parameters with fatality risk.

The actual-regression results with the new database in Table 3-5 are quite close to the previous report's "lower-estimate scenario" that attributed to 100-pound mass reductions a 1.02 percent fatality increase in the lighter cars, a 0.73 percent societal benefit in the heavier LTVs, and increases of 0.44 and 0.41 percent in the heavier cars and lighter LTVs. The lower-estimate scenario was an attempt to override, based on judgment of NHTSA staff, what were perceived as excessive influences of lighter/smaller cars being poorly driven and possible distortion of the effects, due to combining disparate types of LTVs, in some of the LTV regressions.[88]

Sensitivity test 1: Models essentially similar to the previous report were applied to the new database to help illustrate how results have changed from the previous report. The same four vehicle classes and six crash types were defined as in the previous report (i.e., with CUVs and minivans grouped with LTVs) and baseline fatalities were not adjusted for the future effect of ESC. The model was "refreshed" only in the use of the new VMT data (because the NASS-

---

[88] Kahane (2010), pp. 517-520 and 531-534.

based estimates of VMT variation by vehicle class in MY 1991-1999 do not extend to MY 2000-2007) and by adding the control variables for ESC, curtain/side air bags, and blocker beams (which are present in many MY 2000-2007 vehicles but in few MY 1991-1999 vehicles) while deleting the variable for frontal air bags (which is present in all MY 2000-2007 vehicles, but not all MY 1991-1999). Table 3-6 shows the estimated effects of 100-pound mass reduction while holding footprint constant, in each type of crash and overall:

TABLE 3-6: ESTIMATED EFFECTS WITH MY 2000-2007 DATABASE
BUT VEHICLE CLASSES AND CRASH TYPES AS IN THE MY 1991-1999 ANALYSIS

FATALITY INCREASE PER 100-POUND MASS REDUCTION (%)

CRASH TYPE	CARS < 2,950	CARS ≥ 2,950	LTVs < 3,870	LTVs ≥ 3,870
1st-EVENT ROLLOVER	- 1.98	- 1.17	- 5.90	- 3.69
HIT FIXED OBJECT	- .98	- .60	- 3.64	- .07
HIT PEDESTRIAN/BIKE/MOTORCYCLE	2.96	.24	- .09	- .65
HIT HEAVY VEHICLE	3.43	2.29	.57	- .68
HIT CAR	.98	1.35	- 2.87	- 1.90
HIT LTV	3.80	3.09	.90	- 1.07
OVERALL	1.36	.95	- 1.94	- 1.34

On the one hand, this hybrid analysis shows the same diminution of the overall effect in the lighter cars (from a 2.21% increase in the 2010 report to 1.36%) and the heavier LTVs (from a 1.90% benefit in the 2010 report to 1.34%). On the other, continuing to combine truck-based LTVs with CUVs and minivans produces strong effects in rollovers and collisions with cars, as in the 2010 report; indeed, the overall effect has diminished despite the continued strong effects in those two types of crashes. The hybrid analysis does not obtain statistically meaningful results for LTVs < 3,870 pounds because there were relatively few such LTVs left in MY 2000-2007, even if CUVs and minivans are included, as curb weights steadily increased.

Sensitivity test 2: A caveat on the confidence bounds in Table 3-5: they estimate the sampling error internal to the regression analyses used in the specific model that generated the point estimates in that table. But that model can be varied by choosing different control variables or redefining the vehicle classes or crash types, for example. The alternative models would produce different point estimates. This can be considered an additional source of uncertainty, but it is more difficult to quantify. NHTSA's 2011 preliminary report stated, "The potential for variation will perhaps be better understood after the public and other agencies have had an opportunity to work with the new database."[89] Indeed, peer-, public-, and government reviewers of the preliminary report tried out or recommended numerous alternative regression models for databases existing at that time. Sections 4.2 and 4.3 of this report update 11 of these analyses with the new databases and present their results as a battery of sensitivity tests of the baseline model.

---

[89] Kahane (2011), p. 81.

DOE suggested one such alternative that was already included the preliminary report: regression analyses that retain police cars, muscle cars, AWD-equipped cars in the car regressions, and full-size vans in the truck-based-LTV regressions, rather than excluding them as the preceding analyses of this chapter. (The CUV/minivan regressions would stay the same.) Including these "niche" vehicles, each with its own pattern of crash types and of relationships with mass and footprint, adds some complexity. It might generate coefficients for mass and footprint that to some extent reflect how the vehicle mix varies for different mass-footprint combinations rather than the underlying relationships of mass and footprint with fatality risk. At a minimum, it behooves to add categorical variables for vehicle type: police car and muscle car in the car regressions, as well as the AWD variable; cargo van and passenger van in the LTV regressions (because cargo and passenger vans have quite different use patterns). This approach, however, does have the plus of using every vehicle case in the databases. Table 3-7 shows the estimated effects of 100-pound mass reduction while holding footprint constant, in each type of crash and overall:

TABLE 3-7: ESTIMATED EFFECTS OF 100-POUND MASS REDUCTION WHILE HOLDING FOOTPRINT CONSTANT
REGRESSIONS INCLUDING POLICE CARS, MUSCLE CARS, AWD-EQUIPPED CARS, AND FULL-SIZE VANS

FATALITY INCREASE PER 100-POUND DOWNSIZING (%)

	CARS		TRUCK-BASED LTVs		CUVs & MINIVANS[90]
	< 3,106	≥ 3,106	< 4,594	≥ 4,594	
1st-EVENT ROLLOVER	- 1.62	- 2.34	.90	- 2.29	- 7.02
HIT FIXED OBJECT	- .32	- 1.37	- 1.19	.28	- 3.61
HIT PEDESTRIAN/BIKE/MOTORCYCLE	2.45	- .09	1.24	- .37	- 1.57
HIT HEAVY VEHICLE	2.50	.65	2.04	- .93	1.94
HIT CAR-CUV-MINIVAN < 3082	.96	.05	- .29	- 1.20	- .09
HIT CAR-CUV-MINIVAN 3082+	.80	2.03	- 1.00	- 1.84	1.68
HIT TRUCK-BASED LTV < 4150	1.72	1.00	- .31	- 1.63	3.82
HIT TRUCK-BASED LTV 4150+	6.28	1.95	4.16	.26	- .93
ALL OTHERS	2.07	.90	.50	- .33	- .39
OVERALL - POINT ESTIMATE	1.81	.49	.49	- .76	- .37
LOWER CONFIDENCE BOUND	.55	- .31	- .40	- 1.40	- 1.55
UPPER CONFIDENCE BOUND	3.07	1.29	1.38	- .12	.81

Including police, muscle, and AWD-equipped cars slightly strengthened the overall coefficient for mass reduction in cars < 3,106 pounds by 0.25 percentage points, from a 1.56 percent fatality increase per 100-pound reduction in Table 3-5 to 1.81 here. The coefficient for cars ≥ 3,106 changed little (.51 to .49). It seems paradoxical that there would be any change in the cars < 3,106 because most police cars, muscle cars and AWD-equipped cars actually weigh more than 3,106 pounds. This may illustrate how the inclusion of "niche" vehicles may distort how the regression allocates effects between mass and footprint; the same thing happened when muscle cars were included in the 2010 report.[91] In the regressions of truck-based LTVs, including full-size vans had an effect of similar size but in the opposite direction, making the overall coefficient

---
[90] Unchanged from Table 3-5.
[91] Kahane (2010), p. 488.

for LTVs ≥ 4,594 pounds 0.42 percentage points more negative, from -.34 to -.76. The coefficient for LTVs < 4,594 pounds stayed nearly the same (.52 to .49). The principal change from Table 3-5 is that the societal benefit of mass reduction in LTVs ≥ 4,594 pounds has become statistically significant (confidence bounds from -1.40% to -.12%). The principal similarity with Table 3-5 is that the overall results remain in the same direction, and the added benefit for mass reduction in the heavier LTVs more or less offsets the added harm in the lighter cars.

3.6 Combined annual effect of mass reduction in several classes of vehicles

Charles Farmer asks in his peer review of NHTSA's 2010 report about the effect of removing 100 pounds from every car and LTV while holding footprint constant; indeed, this is not estimated in the 2010 report (although the 2003 report did estimate the annual effect of removing 100 pounds with commensurate footprint reductions). It is useful to make the question more general: what would be the annual effect on fatalities of removing x pounds from the heavier LTVs, y pounds from the lighter cars, and so on? The issues involved are adjusting the baseline fatality counts to annual levels and addressing the issue of double-counting (namely, when a FARS crash involves two or more MY 2000-2007 vehicles, it will appear multiple times in the vehicle-oriented database).

As in NHTSA's 2003 report, the starting point is FARS for the last five CY in the database – in this case, 2004-2008.[92] There were 207,736 fatalities in crashes during that 5-year period, an average of 41,547 per year. However, quite a few involved only motorcycles, heavy trucks, or other vehicles that are not cars or LTVs; 184,970 of the fatalities occurred in crashes involving at least one car or LTV, an average of 36,994 per year. But these crashes involve an on-road fleet including vehicles of any model year, sometimes long before 2000, and most of the on-road fleet was not yet ESC-equipped. If all the cars and LTVs on the road had already been equipped with ESC, NHTSA estimates that there would have been only 28,078 fatalities per year, not 36,994.[93] This number, 28,078 will serve as the baseline annual fatalities in the post-ESC environment. Furthermore, 17,016 of these 28,078 fatalities would have occurred in crashes involving exactly one light vehicle (i.e., a car or LTV), 9,210 in crashes involving exactly two light vehicles, and 1,852 in crashes involving three or more light vehicles.

Again paralleling the 2003 report, the vehicle sales mix of the four most recent model years in the database, in this case MY 2004-2007 is postulated to continue into the indefinite future until it becomes the entire on-road fleet, having replaced all earlier vehicles. The 28,078 annual fatalities are allocated to vehicle types and crash types based on the experience of MY 2004-2007 vehicles in CY 2004-2008, adjusted for ESC. The 17,016 fatalities in crashes involving one light vehicle would have included, for example, 272 fatalities of cars < 3,106 pounds in first-event rollovers, 323 fatalities of cars ≥ 3,106 pounds in first-event rollovers, and so on.[94] For

---

[92] Kahane (2003), pp. 104-109.
[93] Sivinski (2011) estimates that ESC reduces fatal first-event rollovers by 56 percent in cars and 74 percent in LTVs; fixed-object impacts by 47 percent in cars and 45 percent in LTVs; and other non-pedestrian crashes by 8 percent in both cars and LTVs.
[94] The actual fatalities on the database in crashes involving one car or LTV – and that car or LTV is MY 2004-2007 – are tabulated by vehicle class and crash type. The actual fatality counts for the non-ESC-equipped vehicles are adjusted downward for ESC effectiveness. Each cell in the table is then multiplied by the same constant so that the cells sum up to 17,016.

these 17,016 fatalities, there is no issue of double-counting, as only one light vehicle was involved in the crash.

The case-vehicle involvements in crashes involving exactly two light vehicles and where the case vehicle and the "other" vehicle are both MY 2004-2007 are subdivided into cells by the type of case vehicle (car < 3,106, car $\geq$ 3,106, truck-based LTV < 4,594, truck-based LTV $\geq$ 4,594, CUV/minivan) and the type of crash (predominantly: hit car/CUV/minivan < 3,082, hit car/CUV/minivan $\geq$ 3,082, hit truck-based LTV < 4,150, and hit truck-based LTV $\geq$ 4,150; but also some other types, such as when the crash involved a heavy truck, pedestrian, or motorcycle in addition to the two light vehicles). Each crash will appear twice in the tabulation, once with vehicle no. 1 as the case vehicle, once with vehicle no. 2 as the case vehicle. However, all the cell counts are then multiplied by the same constant so they will add up to 9,210, the annual number of fatalities in such crashes. That addresses the issue of double-counting, for even though the crashes appear twice, the cell counts add up only to 9,210 the number of annual fatalities in the crashes.

However, for the subset of these crashes that involved only the two light vehicles and no other units, the computational model intentionally double-counts because the effects of mass reduction in each of the two light vehicles are additive. These effects are tallied separately and eventually summed. For example, when a car $\geq$ 3,106 pounds collides with a truck-based LTV $\geq$ 4,594 pounds, according to the regressions (Table 3-1), removing 100 pounds from the car would have increased societal risk by 2.34 percent whereas removing 100 pounds from the LTV would have reduced societal risk by 1.37 percent; thus, removing 100 pounds from both would have increased risk by an estimated net 0.97 percent.

The case-vehicle involvements in crashes involving three or more MY 2004-2007 light vehicles are likewise subdivided by the type of case vehicle and the type of crash. Almost all of these are the last type of crash ("all others") and none are of the crash types that involve two light vehicles and nothing else. All the cell counts are then multiplied by the same constant so they will add up to 1,852, the annual number of fatalities in such crashes. Even though some crashes may appear multiple times, the cell counts add up only to 1,852, essentially pro-rating the cases and avoiding double-counting.

Table 3-8 shows how the computational model works to estimate the annual effect of removing 100 pounds from every vehicle. It tabulates a year's crash fatalities by crash type and case-vehicle type. The first column of numbers, "annual crash fatalities after ESC" adds up to exactly 28,078. In the first four crash types and the last one, these are the sums of the FARS variable FATALS for the vehicle records on the database, adjusted downward for ESC effectiveness and multiplied by constants, as described above, in order that the fatalities in crashes involving one light vehicle add up to 17,016 and the fatalities in crashes involving three or more light vehicles add up to 1,852.

TABLE 3-8: ESTIMATED ANNUAL EFFECT OF 100-POUND MASS REDUCTION IN ALL VEHICLES HOLDING FOOTPRINT CONSTANT

CRASH TYPE	VEHICLE TYPE	ANNUAL CRASH FATALS AFTER ESC		FATALITY INCREASE PER 100 LB RED (%)	MASS RED	FATALITY INCREASE
1st-EVENT ROLLOVER	CAR < 3106	272		-1.83	100	- 5.0
	CAR 3106+	323		-2.89	100	- 9.3
	TRUCK-BASED LTV < 4594	213		.66	100	1.4
	TRUCK-BASED LTV 4594+	452		-1.28	100	- 5.8
	CUV OR MINIVAN	131		-7.02	100	- 9.2
HIT FIXED OBJECT	CAR < 3106	1067		- .48	100	- 4.9
	CAR 3106+	1656		-1.29	100	-21.4
	TRUCK-BASED LTV < 4594	592		-1.39	100	- 8.2
	TRUCK-BASED LTV 4594+	1051		.76	100	8.0
	CUV OR MINIVAN	489		-3.61	100	-17.7
HIT PEDESTRIAN/BIKE/MOTORCYCLE	CAR < 3106	1139		2.03	100	23.1
	CAR 3106+	1811		- .14	100	- 2.5
	TRUCK-BASED LTV < 4594	871		1.07	100	9.3
	TRUCK-BASED LTV 4594+	1995		- .05	100	- 1.1
	CUV OR MINIVAN	1105		-1.57	100	-17.4
HIT HEAVY VEHICLE	CAR < 3106	630		2.26	100	14.3
	CAR 3106+	864		.39	100	3.4
	TRUCK-BASED LTV < 4594	365		1.62	100	5.9
	TRUCK-BASED LTV 4594+	638		.32	100	2.0
	CUV OR MINIVAN	365		1.94	100	7.1
HIT CAR-CUV-MINIVAN < 3082	CAR < 3106	161*	327**	.76	100	2.5
	CAR 3106+	300*	721**	.26	100	1.9
	TRUCK-BASED LTV < 4594	186*	394**	- .09	100	- .4
	TRUCK-BASED LTV 4594+	223*	869**	- .91	100	- 7.9
	CUV OR MINIVAN	176*	467**	- .09	100	- .4
HIT CAR-CUV-MINIVAN 3082+	CAR < 3106	723*	1230**	.48	100	5.9
	CAR 3106+	1330*	2512**	1.62	100	40.7
	TRUCK-BASED LTV < 4594	421*	767**	- .71	100	- 5.4
	TRUCK-BASED LTV 4594+	757*	1739**	-1.37	100	-23.9
	CUV OR MINIVAN	465*	1074**	1.68	100	18.1
HIT TRUCK-BASED LTV < 4150	CAR < 3106	129*	247**	1.17	100	2.9
	CAR 3106+	169*	281**	.53	100	1.5
	TRUCK-BASED LTV < 4594	124*	212**	- .63	100	- 1.3
	TRUCK-BASED LTV 4594+	67*	199**	- .96	100	- 1.9
	CUV OR MINIVAN	38*	127**	3.82	100	4.8
HIT TRUCK-BASED LTV 4150+	CAR < 3106	656*	971**	6.06	100	58.8
	CAR 3106+	782*	1389**	2.34	100	32.6
	TRUCK-BASED LTV < 4594	343*	643**	4.46	100	28.7
	TRUCK-BASED LTV 4594+	412*	854**	.53	100	4.6
	CUV OR MINIVAN	277*	576**	- .93	100	- 5.3
ALL OTHERS	CAR < 3106	736		1.95	100	14.3
	CAR 3106+	1268		1.16	100	14.8
	TRUCK-BASED LTV < 4594	493		.73	100	3.6
	TRUCK-BASED LTV 4594+	1172		- .11	100	- 1.3
	CUV OR MINIVAN	642		- .40	100	- 2.5
		======				=======
ALL CRASH TYPES AND ALL VEHICLE TYPES		28078				157.2

* Including each crash only the first time it appears in the database: for tallying annual crash fatalities
** Including each crash both times it appears: for computing effects of mass reduction in each vehicle

Table 3-8 shows two counts for the four types of crashes that involve two light vehicles and no other traffic units. The smaller number on the left only counts a crash the first time it appears on the database (i.e., it does not count subsequent vehicle records with the same CY and ST_CASE) to assure that the entire left column adds up to 28,078. The larger number on the right counts the crash both times it appears (most often in two different rows) to allow tallying the effects of mass reduction in either vehicle. The next column is the percent societal fatality increase per 100-pound mass reduction while holding footprint constant: the same numbers as in Tables 3-1 and 3-5, based on the regression coefficients.

The next column is the amount of mass reduction in each vehicle class. In Table 3-8 it is 100 pounds for all vehicle classes. The final column is the annual societal fatality increase associated with the mass reduction in that class of vehicle in that type of crash. It is the product of the annual crash fatalities (the only number shown for the first four and the last crash type, the larger number on the right for the four crash types involving two light vehicles and no other units), the percent effect per 100 pounds, and the amount of mass reduction (which in Table 3-8 is always 100 pounds).

The last column adds up to an estimated annual increase of 157 fatalities if all vehicles became 100 pounds lighter without changing footprint: an increase of 0.56 percent over the 28,078 annual baseline fatalities after ESC.

Confidence bounds and a t-test for the above point estimate (157) are obtained by the same method as the preceding section. For each one of the 45 regression coefficients shown in the middle column of Table 3-8, there are already 10 alternative pseudo-estimates whose variation characterizes the uncertainty contributed by the FARS data and 11 alternative pseudo-estimates that incorporate the uncertainty contributed by the State data. The overall effect in Table 3-8 is recomputed using, for each of the 45 combinations of vehicle class and crash type, the pseudo-estimate coefficient for the FARS cases with ST_CASE ending in 0 substituted for the point-estimate coefficient (but the same weighting factor); again recomputed using the pseudo-estimates for the FARS cases with ST_CASE ending in 1; and so on. The standard error of the 10 resulting pseudo-estimates of the overall effect serves as the FARS-based component of its sampling error. The overall effect is likewise recomputed 11 times using the pseudo-estimate coefficients for the various State files to estimate the State-based component of its sampling error. The standard deviation of the original point estimate (157) is the root-sum-of-squares of the FARS- and State-based components. The two-sided 95% confidence bounds are the point estimate ±2.262 standard deviations, where 2.262 is the 97.5th percentile of a t-distribution with 9 df.[95] The t-test with 9 df can also be applied to test if the point estimate is statistically significant.

The two-sided 95% confidence bounds are an annual increase of 157 ± 196 fatalities – i.e., they range from a reduction of 39 fatalities to an increase of 353. The point estimate is not a statistically significant increase (t = 1.82, df = 9). (As in Section 3.5, the confidence bounds will exclude zero if and only if the t-test is significant – i.e., |t| > 2.262.) In percentage terms, the

---

[95] The original point estimate, not the average of either of the 10 or of the 11 pseudo-estimates is used as the center of the confidence interval. The pseudo-estimates are only used to compute the standard errors.

composite effect ranges from a 0.14 percent reduction to a 1.26 percent increase (0.559 ± .696%).

The scenario of removing 100 pounds from every vehicle, while useful for illustrative purposes is not likely to happen. A hundred pounds is twice the proportion of the mass of a 2,500-pound car as a 5,000-pound pickup truck or CUV; most scenarios contemplate proportionately greater or at least equal mass reduction from the heavier vehicles. The computational model in Table 3-8 allows mass reduction to vary among the vehicle classes. Here are three hypothetical scenarios that involve removing different but relatively small amounts of mass:

Proportionate reduction: MY 2004-2007 vehicles weigh an average of 3893 pounds. A hundred pounds is 2.57 percent of 3893. The average weights of the five vehicle classes are 2727, 3459, 4078, 5331, and 3938 pounds, respectively; 2.57 percent of these averages are 70, 89, 105, 137, and 101 pounds. Rather than 100 pounds apiece, 70 pounds are removed from cars < 3,106, 89 pounds from cars ≥ 3,106, and so on. Annual fatalities would increase by an estimated 108 ± 196. It is not a statistically significant increase (t = 1.25). The point estimate is smaller than the effect of reducing 100 pounds from every vehicle (+157 fatalities).

More reduction in heavy LTVs and no reduction in the lighter cars: Double the previous scenario's mass reduction in the heavier truck-based LTVs from 137 to 274 pounds; no mass reduction in the cars < 3,106; and the same reductions as the proportionate-reduction scenario in the other three groups. Annual fatalities would be <u>reduced</u> by an estimated 8 ± 263. This is not a statistically significant benefit (t = -.06).

More reduction in heavy LTVs and at least some reduction in all groups: Reduce mass by 14 pounds in the lighter cars (.2 of the 2.57%), by 247 pounds in the heavier truck-based LTVs (1.8 times 2.57%), by 71 pounds in the heavier cars (.8 of 2.57%), by 121 pounds in CUVs and minivans (1.2 times 2.57%) and by 105 pounds in the lighter truck-based LTVs (exactly 2.57%). This scenario is safety-neutral – i.e., the point estimate is zero change in fatalities. The confidence bounds for the change in annual fatalities are an estimated ± 240.

Each of these scenarios involves relatively small amounts of initial mass reduction in the immediate post-ESC environment. These estimates are not intended as substitutes for the Volpe model, which will track the longer-term effects of more extensive mass reduction in stages and by different amounts depending on the make-model. An important feature of the Volpe model absent here is that after successive mass reductions, cars that originally exceeded 3,106 pounds will eventually fall under 3,106 pounds and then each additional pound of mass reduction would have a more harmful effect (namely, the coefficients for the cars < 3,106 pounds); likewise for truck-based LTVs. Instead, the point of these scenarios is to illustrate the ranges of point estimates and confidence bounds that can be obtained for initial mass reduction: no judicious combination of mass reductions in the various classes of vehicles results in a statistically significant fatality increase and many potential combinations are safety-neutral as point estimates.

3.7 Effect of reducing mass and footprint (downsizing)

All of the analyses so far estimated the effect of mass reduction while holding footprint constant, which NHTSA assumes is most likely in the future given the disincentive to shrink footprint due to the footprint-based CAFE standards. It is also possible to estimate the effect of downsizing, namely, reducing mass with historically commensurate reductions in footprint (size) – more exactly, comparing the societal fatality rates of groups of vehicles of the same type (e.g., cars) but different mass and footprint, the heavier vehicles having typically larger footprints than the lighter vehicles. This can be accomplished by simply running the 27 basic regressions without the footprint variable, but all other variables unchanged: the effect of historically commensurate footprint reduction will be an implicit component of the regression coefficient for mass.

Table 3-9 lists the 45 coefficients for curb weight in the regressions that omit the footprint variable, by vehicle type and crash type; the overall effects (point estimates and their confidence bounds) of downsizing by 100 pounds in each of the five classes of vehicles, computed using the same post-ESC baseline fatalities as in Table 3-5; and the composite annual effect (point estimate and its confidence bounds) of downsizing all vehicles by 100 pounds, computed as in Table 3-8.

Downsizing by 100 pounds (which, by definition, includes a historically commensurate reduction in footprint) is associated with substantial fatality increases in passenger cars (often more than 2%), except in their collisions with other cars. Benefits of downsizing exceeding 1 percent are found only in the heavier LTVs' collisions with cars.

TABLE 3-9: ESTIMATED EFFECTS OF 100-POUND MASS REDUCTION <u>WITHOUT</u> HOLDING FOOTPRINT CONSTANT
BY VEHICLE CLASS AND CRASH TYPE (DOWNSIZING)

FATALITY INCREASE PER 100-POUND DOWNSIZING (%)

	CARS		TRUCK-BASED LTVs		CUVs & MINIVANS
	< 3,106	≥ 3,106	< 4,594	≥ 4,594	
1st-EVENT ROLLOVER	2.93	3.68	1.94	- .49	- .49
HIT FIXED OBJECT	1.96	1.94	.88	2.08	.72
HIT PEDESTRIAN/BIKE/MOTORCYCLE	2.59	.58	- .33	- .86	-1.37
HIT HEAVY VEHICLE	4.16	2.71	2.47	.82	4.56
HIT CAR-CUV-MINIVAN < 3082	.90	.44	- .32	-1.05	- .52
HIT CAR-CUV-MINIVAN 3082+	.78	2.00	- .36	-1.17	.43
HIT TRUCK-BASED LTV < 4150	3.68	3.64	.53	- .30	1.62
HIT TRUCK-BASED LTV 4150+	7.19	3.72	2.57	- .58	1.15
ALL OTHERS	2.68	2.06	.24	- .40	1.15
OVERALL - POINT ESTIMATE	2.78	1.97	.47	- .38	.61
LOWER CONFIDENCE BOUND	1.64	.96	.04	-1.00	- .25
UPPER CONFIDENCE BOUND	3.92	2.98	.90	.24	1.47

100-POUND DOWNSIZING IN ALL VEHICLES: FATALITIES INCREASE BY 1.55 ± .66% (436 ± 186 PER YEAR)

Across crash types, societal fatality risk increases by a statistically significant 2.78 ± 1.14 percent for cars < 3,106 pounds (t = 5.53) and by a statistically significant 1.97 ± 1.01 percent for cars ≥ 3,106 pounds (t = 4.43). The fatality increase in truck-based LTVs < 4,594 pounds is smaller, but still a significant .47 ± .43 percent (t = 2.48). There is a non-significant .38 percent societal fatality reduction for downsizing truck-based LTVs ≥ 4,594 pounds (t = -1.39). The effect of downsizing in CUVs and minivans is a non-significant .61 percent increase (t = 1.62) – as opposed to a .37% fatality reduction if 100 pounds are removed while holding footprint constant.

The estimated effect of downsizing all vehicles by 100 pounds is a statistically significant fatality increase of 1.55 ± .66 percent, which would amount to an increase of 436 ± 186 fatalities per year. By contrast, the effect of 100-pound mass reduction while maintaining footprint was a non-significant increase of 157 ± 196 per year. (The estimate for downsizing has slightly narrower confidence bounds because there is no variance inflation from having mass and footprint in the same regression).

As mentioned above, the "footprint-based" CAFE standards are intended to discourage downsizing by setting higher mpg levels for smaller footprints, but it is possible that vehicles could become smaller in the future as well as lighter, on the average, as a result of changing consumer preferences or other factors perhaps not directly related to the CAFE standards. However, the main point of these analyses of downsizing is to illustrate two trends in the data, namely:

- Downsizing continues to have a significant harmful effect in the new database; the results for mass reduction in the preceding sections are largely non-significant because the effect of mass reduction while maintaining footprint, if any, is small, not because the data was insufficient in quantity.
- Nevertheless, the composite effect of downsizing (436 ± 186 added fatalities per year) is not nearly as large as it was in NHTSA's 2003 report, which estimated 1118.[96] This is partly due to the reduction in baseline fatalities, especially as a result of ESC. But it may also be additional evidence that vehicles and perhaps driving patterns changed between MY 1991-1999 and 2000-2007. Many vehicles were discontinued or redesigned: the smallest and lightest vehicles, if they still existed in MY 2000-2007, had larger footprint and more mass; SUVs that once had poor static stability were improved; poor performers on crash tests were redesigned, usually adding mass in the process (see Section 1.4). The tendency of small, light vehicles to be driven poorly was apparently not as strong as it used to be (see Section 1.7). It will be interesting to see which of these ameliorating trends may continue in the future, especially if a new generation of substantially smaller and lighter vehicles (designed to a high level of safety) is introduced.

---

[96] Kahane (2003), sum of 71 and 234 on p. ix, 216 and 597 on p. xi.

3.8 Regression analyses including IIHS crash-test ratings

The Insurance Institute for Highway Safety's (IIHS) crash-test ratings for frontal-offset and side impacts are potentially useful for explaining some of the differences in fatality rates between makes and models. However, a fair proportion of MY 2000-2007 makes and models were not tested, especially in side impact. For this project, NHTSA reviewed the IIHS results and translated the make-model names to the 5-digit make-model and vehicle-group codes in NHTSA's database. The agency cannot guarantee that the translations are correct and cautions that ratings for specific vehicles should be obtained from the source material (available to the public at www.iihs.org) rather than from NHTSA's database. Table 3-10, based on counts of induced-exposure cases indicates that offset-frontal ratings are extensively available on selected groups of vehicles:

TABLE 3-10: PERCENT OF MY 2000-2007 VEHICLES WITH IIHS RATINGS

	Offset Frontal	Side Impact
2-door cars	41	8
4-door cars	91	35
Compact or 1500-series pickup trucks	98	11
Heavy-duty pickup trucks	none	none
Truck-based SUVs < 4,594 pounds	97	15
Truck-based SUVs ≥ 4,594 pounds	29	7
CUVs	86	46
Minivans	99	13

Side-impact testing was not yet comprehensive in the earlier model years. They are not present on enough vehicles to recommend including them in the current statistical analyses. Frontal-offset testing was more comprehensive, but not consistently available across vehicle groups. It was customary or almost universal on 4-door cars, CUVs and minivans, but not 2-door cars. For truck-based LTVs it was limited primarily to the lighter vehicles and nearly absent on the heaviest pickup trucks and SUVs. This essentially limits the data to exploratory analyses of 4-door cars and CUVs/minivans. There is less potential value in analyzing the truck-based LTVs because (1) there will be little or no data for the heavier vehicles and (2) the ratings are probably less important here because most of the non-rollover fatalities are in the "other" vehicle and will not likely be influenced by the test performance of the case vehicle.

The issue in these analyses is not whether frontal-offset ratings have a correlation with fatality risk, because they do. Farmer's statistical analysis of relative fatality risk in actual crashes between cars with good and poor frontal-offset ratings, controlling for covariates such as the mass of the vehicles and the age of the drivers, shows a 34-percent lower fatality risk for the driver of the good-rated vehicle across all crashes; the reduction is 74 percent in head-on crashes.[97] Statistical analyses of other ratings show similar results. An early study of NHTSA's NCAP program showed a 26-percent fatality reduction for a belted driver in cars with a combination of good test results in head-on collisions with cars that had poor test results.[98] A more recent study of Euro NCAP compared fatality risk in the better-rated and poorer-rated car across all two-vehicle crashes. The reductions were: 68 percent for a 5-star relative to a 2-star car; 35 percent for 4-star relative to 2-star; and 35 percent for 4-or-5 stars relative to 2-or-3 stars.[99]

The issue is whether the inclusion of the ratings in the regression analyses influences the coefficients for mass and/or footprint. That will only happen if the ratings also have a correlation with mass or footprint. In 4-door cars, mass has a significant $r = .31$ correlation with the overall frontal-offset rating (in the direction higher mass $\leftrightarrow$ better rating) and footprint, $r = .30$. Such relationships could raise complicated questions about cause and effect – e.g., if crush space between the windshield and front axle (footprint) is increased to improve test performance, should the fatality reduction be attributed to the rating or to footprint? But the questions are premature; the first task is to see if the inclusion of the ratings actually influences the coefficients. In CUVs and minivans, these correlations are only .15 and -.13, respectively, and the influence is likely to be smaller.

The first exploratory regressions for 4-door cars do not address individual crash types but analyze the fatality rate for all types of crashes where frontal crashworthiness of the case vehicle is potentially important: collisions with objects or other vehicles (except motorcycles) – i.e., excluding first-event rollovers, other non-collisions, crashes involving non-occupants or motorcycles, and single-vehicle crashes where it is not so clear what happened first. The frontal-offset ratings included in the analysis – one at a time – are: overall, structure, head, chest, or kinematics. Two formulations are considered: (1) as a dichotomous variable, with a value of 0 for "good" or "acceptable" performance and 1 for "marginal" or "poor" performance; (2) as a categorical variable, retaining all four possible values, with "good" performance as the default value and thus obtaining separate coefficients for the other three ratings.

Table 3-11 shows the estimated percent fatality increases for 100-pound mass reductions or 1-square-foot footprint reductions and for the less favorable ratings (with chi-square) in the regressions:

---

[97] Farmer, C. M. (2005). "Relationships of Frontal Offset Crash Test Results to Real-World Driver Fatality Ratings," *Traffic Injury Prevention*, Vol. 6, pp. 31-37.
[98] Kahane (1994), p. x.
[99] Kullgren, A., Lie, A., and Tingvall, C. (2010). "Comparison Between Euro NCAP Test Results and Real-World Crash Data," *Traffic Injury Prevention*, Vol. 11, pp. 587-593.

TABLE 3-11: INFLUENCE OF IIHS OFFSET-FRONTAL RATINGS ON MASS AND FOOTPRINT COEFFICIENTS
COLLISIONS OF 4-DOOR CARS WITH OBJECTS OR VEHICLES (OTHER THAN MOTORCYCLES)
EXCLUDING POLICE CARS AND CARS WITH ALL-WHEEL DRIVE

	FATALITY INCREASE (%) PER				
	100-POUND MASS REDUCTION		SQ FT FOOTPRINT	FATALITY INCREASE (%)	
REGRESSION	< 3106	≥ 3106	REDUCTION	FOR M-OR-P	CHI-SQUARE
WITHOUT IIHS RATINGS					
ALL CARS	1.59	.48	1.99		
ALL CARS WITH IIHS RATINGS	1.77	1.24	1.75		
WITH DICHOTOMOUS RATING (G/A,M/P)					
OVERALL FRONTAL	1.62	.93	1.66	10	36.9
STRUCTURE	1.78	.94	1.62	10	37.1
HEAD	1.77	1.37	1.60	16	19.7
CHEST	N/A BECAUSE ALL MY 2000-2007 CARS ARE GOOD OR ACCEPTABLE				
KINEMATICS	1.64	1.29	1.58	6	20.7
WITH CATEGORICAL RATING (G,A,M,P)				FOR A,M,P	MAX CHI^2
OVERALL FRONTAL	1.49	1.02	1.82	4,21, 7	58.7
STRUCTURE	1.70	.94	1.59	0, 8,14	20.3
HEAD	1.86	1.45	1.43	2, 1,18	23.4
CHEST	1.58	1.24	1.81	6,N/A,N/A	2.5
KINEMATICS	1.63	1.73	1.59	8,11, 4	32.2

The first regression includes all 4-door cars in the database (except police cars and those with AWD) and none of the IIHS ratings. Fatality risk increased by an estimated 1.59 percent per 100-pound reduction in the lighter cars, 0.48 percent in the heavier cars, and also increased 1.99 percent per square foot of footprint reduction in all cars. The second regression is limited to the cars for which IIHS ratings are available, but the ratings are not yet variables in the regression. This is the true starting point for the exploratory analysis, because all the subsequent regressions are for the same cars. If the cars with ratings (91% of all 4-door cars according to Table 3-10) were a random subset, the coefficients should be about the same as in the preceding regression. But the effect of mass reduction in the heavier cars went up from 0.48 to 1.24 percent, which turns out to be a larger influence than any of the IIHS-rating variables. It suggests the cars that have ratings might not be that representative a subgroup of the heavier cars, a possible caveat on the analyses.

The third regression is identical to the second, except it includes the "overall frontal-offset" rating, expressed as a dichotomy (good-or-acceptable versus marginal-or-poor). A marginal-or-poor rating is associated with a quite significant 10 percent fatality increase ($\chi^2 = 36.9$). (That increase is impressive, considering that many of the fatal crashes hardly resemble the frontal-offset test. For example, over 40% of the impacts are not frontal, over 40% of the victims were unbelted, and 9 percent were not front-seat occupants; in all, only 28 percent of the fatalities were belted, front-seat occupants in frontal or oblique-frontal impacts.[100] Also, unlike the statistical studies cited above, the data is not limited to car-to-car collisions, but includes car-to-LTV and car-to-object.) But the mass coefficient in the lighter vehicles only fell from 1.77 to 1.62, a drop of 0.15 percentage points. It is negligible relative to the sampling-error confidence bounds for these coefficients, which are typically a full percentage point or substantially more. The mass coefficient in the heavier cars fell from 1.24 to 0.93, a somewhat larger drop of 0.31 percentage points but still small relative to sampling error. The footprint coefficient dropped by a mere 0.09 percentage points. When dichotomous variables for the "structure," "head," or "kinematics" ratings are substituted for the overall rating, the changes in the mass and footprint coefficients range from a slight increase to a drop of about the same magnitude as with the "overall" rating – even though all three ratings have statistically significant association with fatality risk.

Replacing the dichotomous rating with the full four categories, as shown in the lower section of Table 3-11, does not add much information or change the results. The mass and footprint coefficients vary within the same ranges as with the dichotomous ratings. None of the ratings shows a steadily increasing risk from good to acceptable to marginal to poor. For example, in the overall rating, the fatality risk for "acceptable" is 4 percent higher than "good"; for "marginal," 21 percent higher than "good"; but for "poor," only 7 percent higher than "good." Such variation is presumably due to limited numbers of make-models in each category and limited crash data.

The "overall" rating expressed as a dichotomy appears to convey the most information in a parsimonious way and it is selected for additional analysis. Table 3-12 identifies the influence of including this rating in the six individual types of crashes where frontal crashworthiness is potentially important.

---

[100] Based on the distribution of MY 2000-2007 car and LTV occupant fatalities in impacts with other vehicles or with fixed objects in CY 2008 FARS.

TABLE 3-12: INFLUENCE OF IIHS OVERALL-FRONTAL RATING (GOOD/AVERAGE VERSUS MARGINAL/POOR) ON MASS AND FOOTPRINT COEFFICIENTS, BY CRASH TYPE
4-DOOR CARS WITH IIHS RATINGS EXCLUDING POLICE CARS AND CARS WITH ALL-WHEEL DRIVE

FATALITY INCREASE (%) PER

CRASH TYPE	100-POUND MASS REDUCTION < 3106	100-POUND MASS REDUCTION ≥ 3106	SQ FT FOOTPRINT REDUCTION	FATALITY INCREASE (%) FOR M-OR-P	CHI-SQUARE
HIT FIXED OBJECT					
REGRESSION W/O RATINGS	.08	.28	2.88		
WITH OVERALL-FRONTAL RATING	-.16	-.20	2.78	13	15.1
HIT HEAVY VEHICLE					
REGRESSION W/O RATINGS	2.85	-.26	3.36		
WITH OVERALL-FRONTAL RATING	2.68	-.63	3.26	11	4.1
HIT CAR-CUV-MINIVAN < 3082					
REGRESSION W/O RATINGS	.74	.14	1.07		
WITH OVERALL-FRONTAL RATING	.63	-.06	1.01	6	1.7
HIT CAR-CUV-MINIVAN 3082+					
REGRESSION W/O RATINGS	.45	2.46	.19		
WITH OVERALL-FRONTAL RATING	.44	2.44	.18	1	.03
HIT TRUCK-BASED LTV < 4150					
REGRESSION W/O RATINGS	1.87	.12	4.14		
WITH OVERALL-FRONTAL RATING	1.56	-.67	3.96	20	14.7
HIT TRUCK-BASED LTV 4150+					
REGRESSION W/O RATINGS	5.89	2.04	2.04		
WITH OVERALL-FRONTAL RATING	5.64	1.36	1.86	19	14.9

In all six types of crashes, fatality risk was higher in the cars with marginal or poor rating; in four types of crashes, it was significantly higher. Including the rating in the regression diminished the fatality increase for mass reduction in the lighter cars (or augmented the benefit) by a range of 0.01 to 0.31 percentage points; in the heavier cars, by a range of 0.02 to 0.79 percentage points. The footprint effect barely diminished by a range of 0.01 to 0.18 percentage points. These influences are consistent with the composite effects in Table 3-11 and they, too, are negligible relative to the sampling error inherent in the coefficients. For example, even the largest influence, 0.79 percentage points in the cars ≥ 3,106 hitting LTVs < 4,150 is small relative to the 95% confidence bounds for that coefficient, which range from -2.53 to +3.59 according to Table 3-5, a width of 6.12 percentage points.[101]

In the composite analysis of CUVs and minivans rated by IIHS, the regression results without an IIHS-rating variable were a -0.11 percent fatality increase per 100 pounds of mass reduction and a 2.39 percent fatality increase per square foot of footprint reduction. With inclusion of the

---

[101] Table 3-5 estimates confidence bounds for the regression including 2-door as well as 4-door cars and not limited to vehicles with IIHS ratings.

overall offset-frontal rating, those became a +0.08 percent fatality increase and a 2.26 percent fatality increase, respectively. In other words, one coefficient changed by a negligible 0.19 percentage points in the direction of greater fatality increase and the other by a negligible 0.13 percentage points in the direction of less fatality increase.

The exploratory analyses generally show that the lower IIHS ratings were associated with higher fatality risk in passenger cars and that including the ratings can influence the mass and footprint coefficients to some extent. However, unlike the significant effects of the ratings on fatality risk itself, the influence of the ratings on the mass and footprint coefficients is small relative to the sampling error of the coefficients. It may be desirable to include the IIHS ratings among the control variables in the basic regressions of future databases when the ratings might become available for all makes and models. But at this time it would not yet seem justifiable to adjust the results in the preceding sections on the basis of the relatively small influences seen in these exploratory analyses of a somewhat limited subsample of the database.

3.9 Overall effect by vehicle class can change if the on-road fleet changes mass

One difference between this analysis and NHTSA's 2010 report is that the overall effect of mass reduction by vehicle class – the "Volpe-model coefficients" – can change if the mass distribution of the on-road fleet shifts, even if there is no shift in the mix of vehicle types. That is because there are two separate crash types, "hit car, CUV or minivan < 3082" and "hit car, CUV or minivan ≥ 3082," with different regression coefficients. If the on-road vehicle fleet of cars, CUVs, and minivans – the candidate partner vehicles – becomes lighter on the average, the computation of the overall effect will give greater weight to the "hit car, CUV or minivan < 3082" coefficient, because there would be relatively more collisions with the lighter vehicles. Similarly, there are two separate crash types, "hit truck-based LTV < 4150" and "hit truck-based LTV ≥ 4150." In particular, when the case vehicle is a car < 3,106 pounds, the mass coefficient in collisions with the lighter LTVs is a 1.17 percent fatality increase per 100-pound reduction in the cars, but in the collisions with the heavier LTVs, the estimated fatality increase is a far more severe 6.06 percent per 100-pound reduction (see Table 3-1). If the on-road fleet of LTVs became substantially lighter, including a greater share of LTVs < 4150 and a smaller share of LTVs ≥ 4150, the 1.17 coefficient would be applied to a larger number of baseline fatalities, and the 6.06 to a smaller number. The Volpe-model coefficient would become smaller (less harm for mass reduction, overall). The 2010 model, by contrast, only had one crash type "hit a car" and one type "hit an LTV" and it would not have been similarly affected by changes in the mass distribution of the partner vehicles.

Nevertheless, these potential changes are unlikely to be of any practical importance in the timeframe of the Volpe model. It will be decades before the on-road fleet becomes substantially lighter than it is now, given the longevity of the vehicles. On the contrary, NHTSA anticipates that the average mass of the on-road fleet will continue to increase for some years because the new-vehicle fleet became heavier during MY 2000-2007. Even as somewhat lighter vehicles enter the fleet starting in 2012, they likely will not initially displace the heavy vehicles of MY 2005-2007 but more likely the vehicles of the 1980s and 1990s that were even lighter than the new vehicles.

Still, it is worthwhile as a sensitivity test to compute the changes in the overall effect of mass reduction by vehicle class in response to hypothetical shifts in the mass of the on-road fleet. Table 3-13 begins with the mass distribution of the truck-based LTVs as it is in the current database and the Volpe-model coefficients as estimated in Table 3-5. The on-road fleet of LTVs is reduced in mass 17 times as follows:

- For LTVs initially ≥ 4,150 pounds, reduce mass by 2 percent of the original curb weight each time until reaching 4,150, then by 41.5 pounds at a time
- For LTVs initially < 4,150 pounds, reduce mass by 1 percent of the original curb weight each time

Table 3-13 shows the average curb weight of the on-road fleet of truck-based LTVs at each stage, gradually falling from 4,304 to 3,349 pounds and the Volpe-model coefficients for the five classes of case vehicles:

TABLE 3-13: ESTIMATED OVERALL EFFECTS OF 100-POUND MASS REDUCTION WHILE HOLDING FOOTPRINT CONSTANT, BY VEHICLE CLASS: CHANGES IN RESPONSE TO A LIGHTER ON-ROAD FLEET OF TRUCK-BASED LTVs

	OVERALL FATALITY INCREASE (%) PER 100-POUND MASS REDUCTION IN THE CASE VEHICLES				
	CARS		TRUCK-BASED LTVs		
AVERAGE WEIGHT OF THE 'OTHER' TRUCK-BASED LTV	< 3,106	≥ 3,106	< 4,594	≥ 4,594	CUVs & MINIVANS
4304 (CURRENT DATABASE)	1.56	.51	.52	- .34	- .38
4236	1.54	.50	.50	- .35	- .37
4170	1.51	.49	.46	- .36	- .35
4105	1.49	.48	.43	- .36	- .34
4042	1.47	.47	.41	- .37	- .32
3980	1.45	.47	.39	- .37	- .30
3921	1.42	.46	.37	- .38	- .28
3862	1.39	.45	.35	- .38	- .26
3806	1.36	.44	.33	- .39	- .24
3752	1.34	.43	.32	- .39	- .23
3700	1.31	.42	.29	- .40	- .20
3650	1.28	.41	.28	- .40	- .19
3603	1.25	.40	.26	- .41	- .16
3558	1.23	.39	.24	- .41	- .14
3514	1.21	.39	.22	- .42	- .12
3471	1.19	.38	.21	- .42	- .11
3429	1.18	.38	.20	- .42	- .10
3389	1.17	.38	.19	- .42	- .09
3349	1.16	.37	.19	- .42	- .08

Cars < 3,106 pounds showed the largest change in the Volpe-model coefficient, diminishing from 1.56 initially to 1.16. That is because these case vehicles have by far the largest difference between the "hit truck-based LTV < 4150" and "hit truck-based LTV ≥ 4150" coefficients. The next three classes of vehicles showed a small diminution of the fatality increase or a small augmentation of the fatality-reducing benefit. For CUVs and minivans, the trend is in the opposite direction because the estimated "hit truck-based LTV ≥ 4150" coefficient is actually more favorable than the "hit truck-based LTV < 4150" coefficient; the limited data for CUVs and minivans results in large sampling errors for the coefficients of individual crash types (± 5.00 percentage points in the collisions with LTVs according to Table 3-5), making it impossible to accurately estimate an effect on the order of 0.50 percentage points or less. One reason that the influence of the mass of the on-road fleet is relatively small is because it only alters the baseline fatalities for the collisions with LTVs. Other types of crashes, such as rollovers, fixed-object, pedestrian, and heavy-truck, would continue to make the same contributions to the Volpe coefficient, because their numbers of baseline fatalities are not influenced by the mass distribution of the on-road light-vehicle fleet.

A sensitivity test similar to Table 3-13, but reducing the mass of the on-road fleet of cars, CUVs, and minivans from the initial average of 3,103 pounds to 2,453 pounds, while leaving the truck-based LTVs unchanged did not influence any of the Volpe coefficients by more than 0.22 percentage points. That is because the difference between the "hit car, CUV or minivan < 3082" and "hit car, CUV or minivan ≥ 3082" coefficients is never as large as the corresponding differences in collisions with LTVs.

One relevant feature of the Volpe model is that after successive mass reductions, cars that originally exceeded 3,106 pounds will at some point fall below 3,106 pounds and then each additional pound of mass reduction would have a more harmful effect (namely, the coefficients for the cars < 3,106 pounds); likewise for truck-based LTVs, the initial incremental benefit of mass reduction in trucks exceeding 4,594 pounds would at some point change to an fatality increase after the trucks fall below 4,594 pounds. The computations in Table 3-13 suggest that this initially unfavorable trend in the Volpe model might eventually be mitigated to some extent by an amelioration of the Volpe-model coefficients as the on-road fleet eventually catches up with the mass reduction in the new vehicles. However, this would not likely be a significant factor in the timeframe of the current regulatory analysis or within the "shelf life" of any of the coefficients estimated in this report.

3.10   Effect of mass reduction for drivers with BAC < .08 or with BAC = 0

Paul Green asked in his peer review if the deletion of alcohol-related crashes from the FARS data would change any results. The principal goal of NHTSA evaluations of vehicle safety is to estimate the societal effect on the entire public without excluding behavior-defined groups. However, analyses limited to drivers with blood alcohol concentration (BAC) < .08 (not impaired) or, alternatively, BAC = 0 could be considered sensitivity tests. Specifically, for these analysis, crash involvements on FARS where the median of the driver's ten actual or imputed values of BAC was .08 or higher (or alternatively: .01 or higher) were deleted. No cases were deleted from the induced-exposure file, where BAC would usually be unreported for these mostly nonfatal crashes. (For the same reason, BAC would also be an unsatisfactory control variable in regression analyses: it is usually unreported or said to be zero for the induced-exposure cases.)

One effect may readily be predicted: drinking drivers account for a large proportion of rollovers and impacts with fixed objects, where mass reduction is usually beneficial; excluding them should reduce the share of rollovers and fixed-object impacts in the baseline fatalities and the overall average effect of mass reduction should become more harmful. On the other hand, there are no obvious reasons why excluding the drinking drivers should affect the coefficients for the individual crash types. Table 3-14 compares the individual and overall mass effects for all drivers (on the left, copied from Table 3-5), for drivers with BAC < .08 (in the middle), and for drivers with BAC = 0 (on the right).

The "fatals after ESC" columns (i.e., the numbers of baseline fatalities that would have occurred if all vehicles had been ESC-equipped) indicate that close to half the rollovers and fixed-object impacts involved drivers with BAC $\geq$ .08 – i.e., the baseline fatalities for the drivers with BAC < .08 is about half as large as the "all drivers" baseline. In all the other crash types, only about 10 percent of the drivers had BAC $\geq$ .08. Because rollover and fixed-object are the two crash types where mass reduction is beneficial, the overall effect of 100-pound mass reduction is generally more harmful for the drivers with BAC < .08 – e.g., a 2.20 fatality increase in cars < 3,106 pounds, as compared to a 1.56 percent increase for all drivers. The only exception is truck-based SUVs $\geq$ 4,594 pounds, where the lower share of rollovers and fixed-object impacts is offset by individual regression coefficients more favorable to mass reduction.

On the other hand, there is no obvious directional change in the coefficients for individual crash types. Of the 45 coefficients, 28 went in the direction of more harm (or less benefit) for mass reduction when the data were limited to drivers with BAC < .08: not significantly more than half. It is perhaps noteworthy that all five coefficients for collisions with heavy trucks became stronger for BAC < .08 and so did the coefficients for passenger cars and LTVs < 4,594 pounds when they hit heavy LTVs, whereas all three coefficients for LTVs $\geq$ 4,594 pounds hitting cars or light LTVs became stronger in the opposite direction (greater benefit for mass reduction in the heavy vehicle). The direction in all these cases is what would be expected based on momentum considerations: net harm for removing mass from the lighter vehicle, net benefit from the heavier vehicle. The drivers with BAC < .08, whose crashes are relatively less often preceded by loss of control, may exhibit more strongly the trends associated with momentum considerations. The crashes of drivers with BAC $\geq$ .08 are probably more often preceded by loss of control, even when they hit other vehicles rather than fixed objects; the benefit of the lighter vehicle being somewhat easier to control might, in these cases, be relatively more important than the momentum considerations.

These trends generally became slightly stronger when the relatively small number of cases with BAC .01-.07 (alcohol-related, but below the legal limit) were also excluded. The overall effect of 100-pound mass reduction for drivers with BAC = 0 becomes a 2.32 fatality increase in cars < 3,106 pounds, as compared to a 1.56 percent increase for all drivers and 2.20 percent for drivers with BAC < .08. The overall effect changes in the same direction and magnitude for cars $\geq$ 3,106 pounds, truck-based LTVs < 4,594 pounds, and CUVs/minivans. Again, the only exception is truck-based LTVs $\geq$ 4,594 pounds, where the overall effect of a 100-pound mass reduction becomes slightly more beneficial, a 0.58 percent fatality reduction.

TABLE 3-14: ALL DRIVERS VERSUS DRIVERS WITH BAC < .08 OR BAC = 0
ESTIMATED EFFECTS OF 100-POUND MASS REDUCTION WHILE HOLDING FOOTPRINT CONSTANT, BY VEHICLE CLASS AND CRASH TYPE

CRASH TYPE	ALL DRIVERS			DRIVERS WITH BAC < .08			DRIVERS WITH BAC = 0		
	FATALS AFTER ESC	FATALITY INCREASE N	%	FATALS AFTER ESC	FATALITY INCREASE N	%	FATALS AFTER ESC	FATALITY INCREASE N	%
				CARS < 3,106 POUNDS					
1st-EVENT ROLLOVER	207	-4	-1.83	116	-2	-1.39	105	-1	-1.32
HIT FIXED OBJECT	813	-4	-.46	485	1	.20	418	3	.65
HIT PEDESTRIAN/BIKE/MOTORCYCLE	871	18	2.03	815	19	2.31	790	17	2.16
HIT HEAVY VEHICLE	471	11	2.26	410	15	3.59	389	18	4.63
HIT CAR-CUV-MINIVAN < 3082	478	4	.76	415	8	1.97	394	7	1.75
HIT CAR-CUV-MINIVAN 3082+	674	3	.48	602	2	.40	569	1	.26
HIT TRUCK-BASED LTV < 4150	351	4	1.17	305	5	1.76	292	6	1.95
HIT TRUCK-BASED LTV 4150+	505	31	6.06	455	31	6.88	437	29	6.66
ALL OTHERS	1,530	30	1.95	1,440	31	2.17	1,394	32	2.26
OVERALL	5,901	92	1.56	5,044	111	2.20	4,788	111	2.32
				CARS ≥ 3,106 POUNDS					
1st-EVENT ROLLOVER	247	-7	-2.89	128	-4	-3.24	107	-2	-1.82
HIT FIXED OBJECT	1,263	-16	-1.29	705	-6	-.90	611	-3	-.48
HIT PEDESTRIAN/BIKE/MOTORCYCLE	1,388	-2	-.14	1,291	6	.48	1,240	7	.55
HIT HEAVY VEHICLE	687	3	.39	593	3	.43	572	6	1.07
HIT CAR-CUV-MINIVAN < 3082	921	2	.26	801	6	.73	767	3	.41
HIT CAR-CUV-MINIVAN 3082+	1,172	19	1.62	1,052	20	1.89	1,012	18	1.78
HIT TRUCK-BASED LTV < 4150	543	3	.53	463	2	.52	444	0	.05
HIT TRUCK-BASED LTV 4150+	727	17	2.34	641	18	2.86	610	18	2.87
ALL OTHERS	2,552	30	1.16	2,360	35	1.47	2,287	34	1.49
OVERALL	9,499	48	.51	8,035	79	.98	7,651	81	1.06

	ALL DRIVERS			DRIVERS WITH BAC < .08			DRIVERS WITH BAC = 0		
	FATALS AFTER ESC	FATALITY INCREASE N	%	FATALS AFTER ESC	FATALITY INCREASE N	%	FATALS AFTER ESC	FATALITY INCREASE N	%
CRASH TYPE									
				PICKUPS & TRUCK-BASED SUVs < 4,594 POUNDS					
1st-EVENT ROLLOVER	162	1	.66	101	1	1.27	93	1	.79
HIT FIXED OBJECT	451	-6	-1.39	258	-1	-.38	229	-0	-.04
HIT PEDESTRIAN/BIKE/MOTORCYCLE	676	7	1.07	625	7	1.15	603	8	1.38
HIT HEAVY VEHICLE	287	5	1.62	250	6	2.25	236	6	2.51
HIT CAR-CUV-MINIVAN < 3082	530	-0	-.09	462	-1	-.30	452	-1	-.26
HIT CAR-CUV-MINIVAN 3082+	485	-3	-.71	418	-1	-.21	400	-1	-.26
HIT TRUCK-BASED LTV < 4150	252	-2	-.63	217	-2	-.76	207	-2	-1.05
HIT TRUCK-BASED LTV 4150+	289	13	4.46	262	13	4.85	256	12	4.74
ALL OTHERS	1,126	8	.73	1,032	7	.63	1,014	7	.73
OVERALL	4,258	22	.52	3,624	28	.77	3,491	30	.86
				PICKUPS & TRUCK-BASED SUVs ≥ 4,594 POUNDS					
1st-EVENT ROLLOVER	345	-4	-1.28	206	-6	-2.80	190	-6	-3.27
HIT FIXED OBJECT	802	6	.76	419	-0	-.05	367	0	.07
HIT PEDESTRIAN/BIKE/MOTORCYCLE	1,516	-1	-.06	1,400	-0	-.04	1,341	-2	-.17
HIT HEAVY VEHICLE	492	2	.32	443	3	.67	423	2	.46
HIT CAR-CUV-MINIVAN < 3082	1,262	-12	-.91	1,134	-12	-1.02	1,103	-11	-.99
HIT CAR-CUV-MINIVAN 3082+	1,155	-16	-1.37	1,028	-16	-1.57	994	-15	-1.49
HIT TRUCK-BASED LTV < 4150	578	-6	-.96	532	-7	-1.30	511	-8	-1.62
HIT TRUCK-BASED LTV 4150+	490	3	.53	442	3	.70	427	3	.61
ALL OTHERS	2,262	-3	-.11	2,087	-4	-.18	2,030	-5	-.25
OVERALL	8,902	-30	-.34	7,692	-39	-.51	7,387	-43	-.58

	ALL DRIVERS			DRIVERS WITH BAC < .08			DRIVERS WITH BAC = 0		
	FATALS AFTER ESC	FATALITY INCREASE N	%	FATALS AFTER ESC	FATALITY INCREASE N	%	FATALS AFTER ESC	FATALITY INCREASE N	%
CRASH TYPE									
				CUVs & MINIVANS					
1st-EVENT ROLLOVER	100	-7	-7.02	68	-6	-8.33	67	-7	-9.98
HIT FIXED OBJECT	373	-13	-3.61	272	-11	-3.91	259	-10	-3.95
HIT PEDESTRIAN/BIKE/MOTORCYCLE	812	-13	-1.57	773	-15	-2.00	756	-15	-2.03
HIT HEAVY VEHICLE	297	6	1.94	275	7	2.68	267	10	3.69
HIT CAR-CUV-MINIVAN < 3082	503	-0	-.09	469	-2	-.38	456	-4	-.83
HIT CAR-CUV-MINIVAN 3082+	569	10	1.68	532	11	2.01	522	10	1.89
HIT TRUCK-BASED LTV < 4150	244	9	3.82	225	10	4.58	219	11	4.92
HIT TRUCK-BASED LTV 4150+	294	-3	-.93	273	-4	-1.36	268	-3	-.99
ALL OTHERS	1,380	-5	-.40	1,332	-2	-.14	1,291	-0	-.03
OVERALL	4,571	-17	-.37	4,219	-11	-.26	4,105	-8	-.19

4. Sensitivity Tests and Response to Comments on the Preliminary Report

4.0 Summary

NHTSA's baseline analysis, in addition to sampling error, has another source of uncertainty, namely that the baseline statistical model can be varied by choosing different control variables or redefining the vehicle classes or crash types. Alternative models could produce different Volpe coefficients. NHTSA garnered 11 plausible alternative techniques that could be construed as sensitivity tests from peer-, public-, and government reviews of the preliminary report. The tests illustrate both the fragility and the robustness of the baseline estimates. On the one hand, the variation among the Volpe coefficients is large relative to the baseline estimate: a range of point estimates similar to the sampling-error confidence bounds of the baseline estimate. On the other hand, the variations are not large in absolute terms. In the alternative models, as in the baseline models, mass reduction tends to be relatively more harmful in the lighter vehicles, more beneficial in the heavier vehicles. The societal effect of mass reduction remains small in the various tests; a judicious combination of mass reductions that maintain footprint and are proportionately higher in the heavier vehicles is unlikely to have a societal effect large enough to be detected by statistical analyses of crash data. This chapter also addresses reviewers' comments on: (1) effect of mass reduction on occupants' fatality risk in the "case" versus the "other" vehicle, (2) indexing footprint to mass or vice-versa, (3) comparing the order of magnitude of the societal effect of mass reduction to the effectiveness of various safety technologies, and (4) parsing the effect on societal fatalities per VMT into effects on fatalities per reported crash and reported crashes per VMT.

4.1 Comments on NHTSA's 2011 preliminary report

NHTSA issued its preliminary report[102] on relationships between fatality risk, mass, and footprint in MY 2000-2007 vehicles for public comment on November 28, 2011.[103] NHTSA requested peer reviews, following OMB guidelines, from the same three researchers who reviewed NHTSA's 2010 report.[104] As of April 2012, there are nine public documents from seven commenters that specifically present and/or recommend analyses as additions or alternatives to those in the preliminary report. They include the three peer reviews:

- Charles M. Farmer, Ph.D., Director of Statistical Services, Insurance Institute for Highway Safety, Arlington, VA[105]

- Paul E. Green, Ph.D., Assistant Research Scientist, Vehicle Safety Analytics, University of Michigan Transportation Research Institute, Ann Arbor, MI[106]

- Mr. Anders Lie, Specialist, Traffic Safety Division, Swedish Transport Administration, Borlange, Sweden[107]

---

[102] Kahane (2011), item 0023 in Docket No. NHTSA-2010-0152, which may be accessed by entering "NHTSA-2010-0152" at http://www.regulations.gov and clicking on "Search."
[103] 76 Fed. Reg. 73008 (November 28, 2011), item 0027 in Docket No. NHTSA-2010-0152.
[104] Instructions for the peer review are items 0024 and 0025 in Docket No. NHTSA-2010-0152.
[105] His peer review is item 0036 in Docket No. NHTSA-2010-0152.
[106] His peer review is item 0037 in Docket No. NHTSA-2010-0152.

Three documents from one commenter in NHTSA's public docket on relationships between fatality risk, mass, and footprint (NHTSA-2010-0152):

- Van Auken and Zellner of Dynamic Research, Inc. (DRI) wrote reports that extensively analyze the databases of NHTSA's preliminary report and recommend additional or alternative analyses.[108]

One comment in NHTSA's public docket on MY 2017-2025 CAFE (NHTSA-2010-0131):

- The International Council on Clean Transportation (ICCT), who sponsored DRI's analyses, incorporated DRI's recommendations into their own comments on the CAFE NPRM.[109]

A report sponsored by DOE, one of NHTSA's partners in CAFE rulemaking; the report is available to the public from NHTSA's docket (NHTSA-2010-0152), among other locations:

- Wenzel of the Lawrence Berkeley National Laboratory (LBNL) performed an assessment of NHTSA's preliminary report. The assessment, dated November 2011, analyzes the databases of NHTSA's preliminary report and demonstrates numerous potential alternative ways to look at the data.[110] (In 2012, Wenzel updated his report with NHTSA's latest databases, additional analyses, and responses to peer-review comments.[111])

A compilation of peer reviews of the preceding report, sponsored by EPA, NHTSA's other partner in CAFE rulemaking; it is available to the public from NHTSA's docket (NHTSA-2010-0152):

- Farmer and Van Auken, Donna Chen and Kara Kockelman (University of Texas), and David Greene (Oak Ridge National Laboratory) reviewed several studies by Wenzel.[112] In preparing its own final report, NHTSA has considered their comments on Wenzel's analyses of NHTSA's database.

---

[107] His peer review is item 0035 in Docket No. NHTSA-2010-0152.

[108] Van Auken, R.M., and Zellner, J. W. (2012b and 2012c). *Updated Analysis of the Effects of Passenger Vehicle Size and Weight on Safety, Phase II; Preliminary Analysis Based on 2002 to 2008 Calendar Year Data for 2000 to 2007 Model Year Light Passenger Vehicles to Induced-Exposure and Vehicle Size Variables*. Report No. DRI-TR-12-01. (Docket Nos. NHTSA-2010-0152-0032 and NHTSA-2010-0152-0033). Torrance, CA: Dynamic Research, Inc.; Van Auken, R.M., and Zellner, J. W. (2012d). *Updated Analysis of the Effects of Passenger Vehicle Size and Weight on Safety; Sensitivity of the Estimates for 2002 to 2008 Calendar Year Data for 2000 to 2007 Model Year Light Passenger Vehicles to Induced-Exposure and Vehicle Size Variables*. Report No. DRI-TR-12-03. (Docket No. NHTSA-2010-0152-0034). Torrance, CA: Dynamic Research, Inc.

[109] Item 0258 in Docket No. NHTSA-2010-0131.

[110] Wenzel (2011).

[111] Wenzel, T. (2012). *Assessment of NHTSA's Report "Relationships Between Fatality Risk, Mass, and Footprint in Model Year 2000-2007 Passenger Cars and LTVs – Final Report."* (To appear in Docket No. NHTSA-2010-0152). Berkeley, CA: Lawrence Berkeley National Laboratory.

[112] Menard, (2012).

4.2 Sensitivity test results

NHTSA's 2011 preliminary report estimated and Chapter 3 of this report updates sampling-error confidence bounds for the baseline statistical model's point estimates of mass reduction: the statistical uncertainty that is a consequence of having less than a census of data. The preliminary report acknowledged another source of uncertainty, namely that the baseline statistical model can be varied by choosing different control variables or redefining the vehicle classes or crash types, for example. Alternative models produce different point estimates. NHTSA believed it was premature to address that in the preliminary report. "The potential for variation will perhaps be better understood after the public and other agencies have had an opportunity to work with the new database."[113] Indeed, the seven commenters listed in Section 4.1 try out or recommend numerous alternative statistical models. From these, NHTSA garnered 11 techniques that could be construed as sensitivity tests of the baseline model, in that they share many features of the baseline model but differ in one or more terms or assumptions. The shared features are:

- Use of NHTSA's fatal-crash and induced-exposure databases, or subsets thereof.
- Regression analyses of societal fatalities per billion VMT, by case-vehicle mass, size (usually footprint but possibly track width and wheelbase), and control variables.
- Separate regressions for different types of case vehicles in different types of crashes (not necessarily the same definitions as baseline), and a weighted average of the effects in a post-ESC vehicle fleet.
- They generate five "Volpe-model coefficients."

NHTSA has applied each of these 11 techniques to the updated databases created for this report to generate alternative Volpe-model coefficients as well as estimates of the annual effect on fatalities of removing 100 pounds from every vehicle, or removing different amounts of mass from the various vehicle types. The range of estimates produced by the sensitivity tests gives an idea of the uncertainty inherent in the formulation of the models. The impact of various assumptions on specific results for certain vehicle or crash types provides insight on relationships between mass or size and fatality risk and on strengths and weaknesses of the modeling approach. However, in presenting this range, NHTSA adds the following caveats:

- The 11 alternatives are, of course, not an exhaustive list of conceivable alternatives. (For example, Wenzel recently added a model that includes the vehicle's purchase price and another model with the median owner income for each make and model.[114]) Yet other techniques may be devised and they could extend the range in either direction.
- The 11 alternatives are inspired by commenters' work but NHTSA's approach and detailed SAS® code are not necessarily identical to theirs. (For example, the tests that exclude drinking drivers originated with Wenzel, but NHTSA excludes all drivers with imputed BAC > 0, whereas Wenzel does not necessarily exclude them.)
- The various alternatives are not all equally plausible or realistic. But how plausible is a judgment call. Section 4.3 discusses what NHTSA perceives as strengths or weaknesses of each analysis. In any case, the range of results is not to be interpreted like a histogram

---

[113] Kahane (2011), p. 81.
[114] Wenzel (2012), Table ES.2, p. ix.

of a normal distribution, where each result has known probability of occurrence, with peak likelihood for the middle result and tailing off to both sides.
- The tables will show only point estimates for the alternative models. In fact, they too, like the baseline estimates, have sampling error (not computed – and not necessarily the same as the baseline). NHTSA has not attempted to define a "composite" of the sampling error and the variation of the point estimates.
- As stated above, the sensitivity testing is limited to models that at least keep the framework of the baseline model (NHTSA's data, regression, fatalities per VMT, etc.).

Table 4-1 estimates the five Volpe coefficients for the baseline model (point estimates and confidence bounds) and the 11 alternative models (point estimates only) – ordered from the lowest to the highest estimated increase in societal risk per 100-pound reduction for cars weighing less than 3,106 pounds. The sources and definitions of the 11 alternative models are as follows:

1. Track width/wheelbase/stopped vehicles: Combines both analysis techniques recommended by Van Auken (2012d) – see next 2 alternative models.

2. With stopped-vehicle State data: Recommended by Van Auken (2012d); induced-exposure cases limited to vehicles that were standing still before the crash, allocating all the registration years and VMT only among these stopped-vehicle cases. NHTSA created a separate databases limited to stopped-vehicle induced-exposure cases, ran the sensitivity tests, and also made the data available to the public at http://www.nhtsa.gov/fuel-economy.

3. By track width and wheelbase: Recommended by Van Auken (2012d); with track width and wheelbase (and curb weight) as independent variables, but not footprint.

4. Without CY control variables: Wenzel (2011), pp. 39-46; baseline regressions but excluding all the CY control variables.

5. CUVs and minivans weighted according to 2010 sales: Recommended by Green in his peer review; CUV and minivan fatality cases and VMT reweighted to reflect relative market shares of CUVs and minivans sold in MY 2010 (i.e., more CUVs, fewer minivans than MY 2000-2007); only CUV/minivan result affected; other results, same as FRIA baseline, shown in *Italics*.

6. Without non-significant control variables: Recommended by Farmer in his peer review; for each of the 27 basic regression analyses, start with the baseline model and delete non-significant ($p > .05$) control variables one-by-one by backward selection.

7. Including muscle/police/AWD cars and full-size vans: Wenzel (2011), p. 48, also Table 3-7 in this report; car regressions include muscle, police, and AWD cars (which were excluded from the PRIA baseline regressions); LTV regressions include full-size vans; *CUV analysis not affected.*

8. Control for vehicle manufacturer: Wenzel (2011), pp. 38-39, modified as recommended by Chen and Kockelman in EPA (2012); baseline regressions with 15 additional control variables denoting the various manufacturers.

TABLE 4-1: BASELINE RESULTS, CONFIDENCE BOUNDS, AND 11 ALTERNATIVE MODELS
Based on Regressions of Societal Fatality Risk per VMT - MY 2000-2007 Cars and LTVs in CY 2002-2008
Fatality Increase (%) Per 100-Pound Mass Reduction While Holding Footprint* Constant

	Cars < 3,106 lbs	Cars ≥ 3,106 lbs	CUVs & Minivans	LTVs[†] < 4,594 lbs	LTVs[†] ≥ 4,594 lbs
FRIA baseline estimate[‡]	**1.56**	**.51**	- .37	**.52**	- .34
95% confidence bounds Lower:	.39	- .59	- 1.55	- .45	- .97
(sampling error) Upper:	2.73	1.60	.81	1.48	.30

ELEVEN ALTERNATIVE MODELS

	Cars < 3,106 lbs	Cars ≥ 3,106 lbs	CUVs & Minivans	LTVs[†] < 4,594 lbs	LTVs[†] ≥ 4,594 lbs
1. Track width/wheelbase/stopped veh	.25	- .89	- .13	- .09	- .97
2. With stopped-vehicle State data	.97	- .62	- .33	.35	- .80
3. By track width & wheelbase	.97	.24	- .24	- .07	- .58
4. W/O CY control variables	1.53	.43	.04	1.20	.30
5. CUVs/minivans weighted 2010 sales	1.56	.51	.53	.52	- .34
6. W/O non-significant control vars	1.64	.68	- .46	.35	- .54
7. Incl. muscle/police/AWD/big van	1.81	.49	- .37	.49	- .76
8. Control for vehicle manufacturer	1.91	.75	1.64	.68	- .13
9. Control for veh manuf/nameplate	2.07	1.82	1.31	.66	- .13
10. Limited to drivers with BAC=0	2.32	1.06	- .19	.86	- .58
11. Limited to good drivers	3.00	1.62	zero	1.09	- .30

*While holding track width and wheelbase constant in alternative model nos. 1 and 3.
[†]Excluding CUVs and minivans.
[‡]Point estimates and confidence bounds from Table 3-5.

9. <u>Control for vehicle manufacturer and nameplate</u>: Wenzel (2011), pp. 38-39, modified as recommended by Chen and Kockelman in EPA (2012); baseline regressions with 20 additional control variables denoting the various manufacturers (treating 5 luxury nameplates as if they were separate "manufacturers").

10. <u>Limited to drivers with BAC=0</u>: Recommended by Green in his peer review of the 2010 report, also Table 3-14 of this report; fatal crash cases limited to case-vehicle drivers with tested or imputed BAC < .01; VMT data same as baseline.

11. <u>Limited to good drivers</u>: Wenzel (2011), pp. 46-47, modified by also excluding imputed BAC ≥ .01; see also Kahane (2003), p. 94; excludes fatal crash cases with BAC > 0, drugs, non-valid license, reckless driving in this crash, and/or history of

multiple crashes or multiple violations during the past 3 years; VMT data same as baseline.

For cars < 3,106, the range of the estimated effects of 100-pound mass reduction in the 11 alternative models is fairly symmetric around the baseline value, a 1.56 percent increase in societal fatalities. The estimates range from a negligible increase of 0.25 percent in the first alternative model up to a 3.00 percent increase in the last model, nearly double the baseline effect. That is more or less the same range as the 95% sampling-error confidence bounds for the baseline estimate: 0.39 to 2.73 percent.

As a general rule, in the alternative models, as in the baseline models, mass reduction tends to be relatively more harmful in the lighter vehicles, more beneficial in the heavier vehicles. Thus, in all models, the point estimate of the Volpe coefficient is positive for cars < 3,106 pounds, and in all models except one, it is negative for LTVs ≥ 4,594 pounds. In fact, the range of alternative estimates for the LTVs ≥ 4,594 pounds, -0.97 to +0.30 percent is exactly the same as the sampling-error confidence bounds.

The models were listed, as stated above, in the order of their Volpe coefficients for cars < 3,106 pounds, from least positive to most positive. As a general rule, within each of the other four vehicle classes, there is also a tendency for the more negative (or less positive) coefficients to be near the top of Table 4-1 and the more positive (or less negative) to be lower in the table.

But here are some exceptions to the general rules of "more positive down the table" and "more negative towards the right of the table." Section 4.3 will discuss them as part of a more in-depth look at the various analysis techniques:

- Using stopped-vehicle instead of induced-exposure cases (model 2) lowers the Volpe coefficient for most of the vehicle classes by about 0.50 percentage points, but it has more impact on the cars ≥ 3,106 pounds, lowering it by 1.33 percentage points.
- Conversely, controlling for manufacturer and nameplate (model 9) has a large increasing effect of 1.31 percentage points on the coefficient for cars ≥ 3,106 pounds.
- Withholding the CY control variables (model 4) leaves the car coefficients unchanged but makes the LTV coefficients substantially more positive. It is the only alternative model that generates a positive effect (i.e., that mass reduction is harmful) for LTVs ≥ 4,594 pounds and also pushes the < 4,594 pound LTV coefficient up to car levels.
- The results for CUVs and minivans are evidently less stable than the others:
  - They are not so symmetric about the baseline effect. In fact, nine of the alternatives show mass reduction to be less beneficial than the baseline effect, or even harmful.
  - The DRI-recommended techniques of using stopped vehicles and/or regression by track width and wheelbase (models 1-3) make mass reduction more beneficial in cars and LTVs, but slightly less beneficial in CUVs and minivans.
  - Control for vehicle manufacturer (model 8) increases the Volpe coefficient for most of the vehicle classes by less than 0.50 percentage points, but it has more impact on the CUVs and minivans, raising it by 2.01 percentage points.
- Models 6, 7, and 10 make mass reduction more harmful than baseline for light cars and more beneficial than baseline for heavy LTVs, at least to some extent.

TABLE 4-2: ESTIMATED ANNUAL EFFECT OF MASS REDUCTION IN ALL VEHICLES (Average: 100 Pounds)
Baseline Results, Confidence Bounds, and 11 Alternative Models
Based on Regressions of Societal Fatality Risk per VMT - MY 2000-2007 Cars and LTVs in CY 2002-2008
Fatality Increase (N and % of annual fatalities) While Holding Footprint* Constant

MASS REDUCTION SCENARIO (Pounds Removed From Each Vehicle Class)

	Scenario 1 (Safety-Neutral)	Scenario 2 (Proportional)	Scenario 3 (100 Pounds)
Cars < 3,106 lbs	14	70	100
Cars ≥ 3,106 lbs	71	89	100
CUVs & Minivans	121	101	100
LTVs[†] < 4,594 lbs	105	105	100
LTVs[†] ≥ 4,594 lbs	247	137	100
Fleet-wide average	100	100	100

ANNUAL FATALITY INCREASE (N and % of 28,078 Baseline Fatalities[‡])

		N	%	N	%	N	%
FRIA baseline estimate		**ZERO**		**108**	**.38**	**157**	**.56**
95% confidence bounds	Lower:	-240	-.85	-88	-.31	-39	-.14
(sampling error)	Upper:	240	.85	304	1.08	353	1.26

ELEVEN ALTERNATIVE MODELS

	N	%	N	%	N	%
1. Track width/wheelbase/stopped veh	-321	-1.14	-235	-.84	-211	-.75
2. With stopped-vehicle State data	-215	-.77	-109	-.39	-72	-.26
3. By track width & wheelbase	-130	-.46	-29	-.10	12	.04
4. W/O CY control variables	151	.54	206	.73	236	.84
5. CUVs/minivans weighted 2010 sales	52	.19	152	.54	200	.71
6. W/O non-significant control vars	-43	-.15	93	.33	153	.54
7. Incl. muscle/police/AWD/big van	-96	-.34	68	.24	138	.49
8. Control for vehicle manufacturer	200	.71	299	1.06	357	1.27
9. Control for veh manuf/nameplate	276	.98	410	1.46	486	1.73
10. Limited to drivers with BAC=0	20	.09	144	.68	201	.94
11. Limited to good drivers	100	.61	190	1.16	235	1.44

*While holding track width and wheelbase constant in alternative model nos. 1 and 3.
[†]Excluding CUVs and minivans.
[‡]There is an average of 28,078 annual fatalities in 2004-2008 in crashes involving at least one car or LTV, adjusted downward for a future all-ESC fleet; different baselines are used for computing percents in alternative model nos. 10 and 11, namely 21,285 in model 10 (annual fatalities in crashes involving drivers with BAC = 0) and 16,340 in model 11 (annual fatalities in crashes involving good drivers).

Table 4-2 estimates the net annual effect on fatalities for the baseline model (point estimates and confidence bounds) and the 11 alternative models (point estimates only) for three scenarios that remove an average of 100 pounds from vehicles while holding footprint constant. The alternative models are listed in the same order as in Table 4-1.

The three scenarios are described in Section 3.6. Scenario 1 is exactly safety-neutral, as a point estimate, in the baseline model: some mass is removed in all vehicles, but disproportionately more in the heavier vehicles and the least in the lighter cars. Scenario 2 is a proportional 2.57% mass reduction, averaging out to 100 pounds over all vehicles, but in absolute terms somewhat higher in the heavier vehicles. Scenario 3 consists of simply removing 100 pounds from all vehicles while holding footprint constant.

Scenario 1 – judicious mass reduction (more in the heavier vehicles and the least in the light cars) that is exactly safety-neutral as a point estimate with the baseline model – generates estimates ranging from an annual savings of 321 lives to an increase of 276 fatalities with the 11 alternative models. Even these extremes are only about 1 percent of annual fatalities in crashes involving cars and LTVs (estimated to be 28,078 per year for an all-ESC fleet), and fairly similar to the sampling-error confidence bounds for the baseline model (± 240); they are absolute numbers that would be difficult to detect by statistical methods, as annual fatalities in the United States typically vary by several hundred or more from year to year.[115] In other words, the sensitivity tests supplement the estimates of sampling error (Section 3.5) in demonstrating that the statistical analyses cannot estimate the effect of mass reduction exactly.

At the same time, the relatively narrow range of the sensitivity tests is quite compatible with this report's conclusion that "the societal effect of mass reduction while maintaining footprint, if any, is small." Specifically, the range, at least in these sensitivity tests, is much smaller than the range of some estimates in analyses of earlier databases, such as a reduction of 1,518 fatalities[116] or an increase of 1,118.[117]

With Scenario 1, five sensitivity tests generate point estimates of a societal benefit for judicious mass reduction, two tests show a negligible increase (0.20% or less) and the baseline analysis is safety-neutral. The four tests that estimate a net increase of 100 or more are: without control for CY (which loses the benefit of mass reduction in the heavier LTVs), controlling for manufacturer (which changes the benefit of mass reduction in CUVs and minivans to harm), controlling for nameplate, and limiting to good drivers (both of which affect all five Volpe coefficients in the direction of more harm/less benefit for mass reduction).

4.3 Discussion of individual sensitivity tests

NHTSA believes each of the 11 sensitivity tests in Tables 4-1 and 4-2 is plausible enough to serve as an alternative estimate for the purpose of assessing the uncertainty of the baseline results and to shed additional light on the relationships between societal fatality risk, vehicle mass, and

---

[115] NHTSA (2010). *Traffic Safety Facts 2009*. Report No. DOT HS 811 402. Washington, DC: National Highway Traffic Safety Administration, Table 1, p. 14, http://www-nrd nhtsa.dot.gov/Pubs/811402.pdf.

[116] Van Auken and Zellner (2005b), for a 100-pound mass reduction while holding track width and wheelbase constant, sum of 836 for passenger cars (Table 2, p. 27) and 682 for LTVs (Table 5, p. 36).

[117] Kahane (2003), for 100-pound downsizing, sum of 71 and 234 on p. ix, 216 and 597 on p. xi.

size. But the agency does not see a compelling reason that any of them should supersede the technique of Chapter 3 as the baseline analysis and the principal estimate. Here is a discussion of the salient features, strengths, and disadvantages of the alternative models. Appendix D lists the estimated regression coefficients for mass and size and their Wald chi-square values for each of the 11 tests.

Induced-exposure crashes limited to stopped vehicles:

	Cars < 3,106 lbs	Cars ≥ 3,106 lbs	CUVs & Minivans	LTVs < 4,594 lbs	LTVs ≥ 4,594 lbs
Baseline	1.56	.51	- .37	.52	- .34
Stopped-vehicle	.97	- .62	- .33	.35	- .80

NHTSA's 1997 report on vehicle size and fatality risk defined induced-exposure crash involvements as "vehicles that had been <u>standing still</u> for some time, for a legitimate reason, and got hit by somebody else. The vehicle should have done nothing to precipitate or contribute to the collision."[118] That is a subset of the involvements more commonly defined as induced exposure, namely all non-culpable vehicles in two-vehicle collisions, regardless of whether they were standing still or moving. There had been early studies that limited induced exposure to standing vehicles, but it was no longer the usual technique.[119] The NAS peer review panel, including D.W. Reinfurt, exposure-data expert, criticized NHTSA's approach: "The [1997] Kahane report does not provide sufficient evidence that the induced exposure group of stopped-vehicle crashes is a suitable surrogate for the vehicle fleet and driving population on the same highways as the fatal crashes."[120]

Van Auken and Zellner's earlier analyses also limited induced exposure to stopped vehicles.[121] However, NHTSA in its 2003 report returned to the "customary approach" of non-culpable vehicles, "whose efficacy is well established."[122] Reinfurt, again a peer reviewer, noted in 2003 that "Induced exposure using the traditional approach of utilizing non-culpable vehicles (drivers) in two vehicle crashes is a large improvement over the 1997 study."[123]

Van Auken subsequently used both methods and found that, given otherwise identical techniques, the regressions on the databases with stopped-vehicle induced exposure usually estimated less harm/more benefit for mass reduction than the other method.[124] In their most recent report, Van Auken and Zellner acknowledge disadvantages of limiting to stopped vehicles, namely: losing nearly ¾ of the crash cases, even more on high-speed roads and in rural

---

[118] Kahane (1997), p. 20.
[119] Haight, F.A. (1973). "Induced Exposure," *Accident Analysis and Prevention*, Vol. 5, pp. 111-126.
[120] NAS (1996). Peer-review letter from D. Warner North to Ricardo Martinez, NHTSA, July 12, 1996, Appendix B. Washington, DC: National Research Council.
[121] Van Auken & Zellner (2003); Van Auken & Zellner (2005a); Van Auken & Zellner (2005b).
[122] Kahane (2003), p. 31.
[123] NHTSA (2003). *Memorandum: Drs. James H. Hedlund, Adrian K. Lund and Donald W. Reinfurt's Reviews and Comments of the Draft Technical Report.* (Docket No. NHTSA-2003-16318-0004). Washington, DC: National Highway Traffic Safety Administration.
[124] Van Auken and Zellner (2012a).

areas, and less acceptance by experts than the other method. But they argue that limiting to stopped vehicles may have a special offsetting benefit in studies of vehicle size and fatality risk:

> "Non-culpable vehicle induced-exposure data can include crashes where the non-culpable vehicle was moving prior to the crash. Therefore, some drivers may be more likely to be involved in these crashes than other drivers, even if the driver is not culpable in the crash. This is because some drivers may be able to avoid a crash in which they are not culpable…due to driver skill, driver alertness and/or ability to properly react in time to avoid a collision…This under-representation in the non-culpable induced-exposure data of good drivers, and over-representation of bad drivers is undesirable…A potential advantage of the stopped-vehicle induced exposure is that it is assumed to be not sensitive to the ability of the driver or vehicle to avoid the crash. This is because the vehicle is stopped and presumably would not have been able to avoid the crash. Therefore this data captures a representative sample of drivers for a given make-model-year vehicle."[125]

Note that the aggregate denominator for societal fatality rates – vehicle registration years and VMT – will not be affected by the choice of induced exposure. A nimble vehicle with low rates of non-culpable crashes will still have the same vehicle years and VMT; it will just allocate these VMT among a smaller number of induced-exposure crash cases. In a regression model without any control variables derived from induced exposure (such as driver age or urbanization), point estimates for the effects of curb weight and footprint will be exactly the same with non-culpable or stopped-vehicle induced exposure. Results will differ only to the extent that the choice of induced exposure changes the distribution of control variables – and then, only if the changes are different for some makes and models than others.

That said, there is evidence to support Van Auken's argument: specifically, the stopped-vehicle subset has a higher proportion of drivers in the 30-50 age range (the most skilled group) and lower proportions of young (inexperienced) and old (often less skilled) drivers than the full non-culpable set. However, there is something else going on that makes both sets of induced-exposure crashes have fewer old drivers than the estimated national distribution of VMT in 2000-2009.

Table 4-3 displays the share of VMT for each five age groups defined in the National Household Transportation Surveys (NHTS) of 2001 and 2009. The NHTS asks drivers how many miles they drive in a year and averages this by age group. These annual-VMT rates are multiplied by the number of licensed drivers to find the total VMT for that age group. The 2001 NHTS rates are multiplied by the 2000 N of licensed drivers, while the 2009 NHTS rates are multiplied by the 2009 N of licensed drivers; the two sets of totals are averaged to obtain an average for the 2000-2009 decade.[126] Table 4-3 also shows the age-group distribution of VMT for MY 2000-2007 cars and LTVs in CY 2002-2008, when the VMT are allocated according to the non-culpable- and according to the stopped-vehicle induced-exposure cases.

---

[125] Van Auken and Zellner (2012d), pp. 20-21.
[126] 2001 and 2009 NHTS average miles per licensed driver, by age group: http://nhts.ornl.gov/2009/pub/stt.pdf, Table 23, p. 43; N of licensed drivers by age group in 2000: http://www.fhwa.dot.gov/ohim/onh00/bar7.htm; N of licensed drivers by age group in 2009: http://www.fhwa.dot.gov/policyinformation/statistics/2009/dl20.cfm.

TABLE 4-3: VMT DISTRIBUTION (%) BY DRIVER AGE GROUP
NHTS VERSUS TWO ESTIMATES BASED ON INDUCED EXPOSURE

Age Group	2000-2009 Average NHTS VMT/Year x N of Drivers	Induced Exposure (MY 2000-07 in CY 2002-08)	
		Non-Culpable	Stopped-Vehicle
19 or younger	2.54	5.53	4.69
20-34	29.35	31.78	30.73
35-54	45.63	42.30	44.88
55-64	13.42	12.17	12.43
65 or older	9.05	8.21	7.27

On the whole, it is remarkable how well both estimates based on induced exposure track the distribution based on NHTS data.[127] Nevertheless, the induced-exposure data overstates the VMT for the youngest group and understates it for the oldest group. To the extent that the stopped-vehicle database has relatively few young and old drivers, it partly remedies the excess of young drivers but makes the shortage of old drivers even worse (only 7.27% age 65+, versus 8.21% with non-culpable induced exposure and 9.05% in NHTS).

Table 4-1 showed that the impact on the Volpe coefficients of limiting the induced exposure to stopped vehicles (test no. 2) was 0.60 percentage points or less for light cars and the two groups of truck-based LTVs and even in the opposite direction for CUVs and minivans – but for cars ≥ 3,106 pounds, it changed the Volpe coefficient from +0.51 in the baseline to -0.62, a change of 1.13 percentage points in the direction of more benefit for mass reduction. This is precisely the group of vehicles with many older drivers. But not in all makes and models. The older drivers are concentrated in the subset of cars that have high mass and not particularly high footprint relative to mass. The regressions based on non-culpable induced exposure say, "These cars with high mass and relatively low footprint have fairly high fatality rates, but they have lots of older drivers, and that explains the high fatality rates." Whereas the regressions based on stopped vehicles say (to a modest extent, but enough to jog the Volpe coefficient), "Older drivers? What older drivers? I don't see that many older drivers. The high fatality rate must be due to higher mass."

Of course, there are other differences between the non-culpable and stopped-vehicle databases – and for those other differences, there is no reference such as NHTS to indicate which of the two distributions is more realistic. In summary, NHTSA considers the stopped-vehicle database to be a plausible alternative model, useful for illustrating the uncertainty of the baseline results. But given the obvious problems with stopped vehicles (loss of ¾ of the cases, even more on high-speed roads, and lack of endorsement from researchers), NHTSA does not believe they should supersede the baseline analysis. The database of stopped vehicles as well as a

---

[127] Three caveats: (1) the NHTS is self-reported annual VMT and may be inaccurate; (2) the induced-exposure data is limited to MY 2000-2007 (relatively new vehicles); (3) the induced-exposure data is based on just 13 States.

corresponding set of fatal crashes is available to the public at http://www.nhtsa.gov/fuel-economy.

Track width and wheelbase rather than footprint:

	Cars < 3,106 lbs	Cars ≥ 3,106 lbs	CUVs & Minivans	LTVs < 4,594 lbs	LTVs ≥ 4,594 lbs
Baseline	1.56	.51	- .37	.52	- .34
By track width & wheelbase	.97	.24	- .24	- .07	- .58

Van Auken and Zellner's earlier analyses include three size-mass variables: curb weight, track width, and wheelbase.[128] NHTSA's 2010 report on the MY 1991-1999 database starts with regressions on those three variables but many subsequent regressions substitute footprint for track width and wheelbase. Footprint "most directly and simply addresses the issue at hand, footprint-based CAFE: what is the historical effect of changing mass given constant footprint?"[129] Furthermore, with that database, many of the regression coefficients for wheelbase were in the perhaps unexpected shorter wheelbase ↔ fewer fatalities direction. Also, the literature suggested that combining parameters – i.e., track width and wheelbase into footprint – is generally advisable for alleviating multicollinearity issues.[130] NHTSA's 2011 preliminary report on the MY 2000-2007 database treated the issue as settled; all its regressions were on curb weight and footprint.

Van Auken's analyses of the MY 2000-2007 database used both sets of size-mass variables and found that, given otherwise identical techniques, the regressions with track width and wheelbase usually estimated less harm/more benefit for mass reduction than regressions with footprint.[131] He suggests that track width and wheelbase are meaningful as separate variables because each has its own natural, physical relationships with certain aspects of crash-proneness and crashworthiness.[132] Some of these natural, cause-and-effect interactions could be lost to the analysis with the more synthetic variable, footprint. Furthermore, if VIF is no greater with the three mass-size variables than with just mass and footprint, there is little added risk of multicollinearity issues.

NHTSA finds that argument sufficiently convincing, at least in theory, to reinstate analyses with track width and wheelbase in this report. It is also true that VIF (measured as the maximum for any of the independent variables in the basic regressions, when curb weight, driver age and CY are entered as simple linear variables) is about the same for track width-wheelbase as for footprint:

---

[128] Van Auken & Zellner (2003); Van Auken & Zellner (2005a); Van Auken & Zellner (2005b).
[129] Kahane (2010), p. 486.
[130] Allison (1999), p. 51; Schadler.
[131] Van Auken and Zellner (2012d), pp. 7-18.
[132] Van Auken and Zellner (2005b), pp. 10-21.

Highest VIF for any variable	Cars	LTVs	CUVs
Curb weight and control variables	2.43	1.86	2.83
Curb weight, footprint, and control variables	7.34	9.80	8.71
Curb weight, track width, wheelbase, & controls	7.65	10.04	7.23

But, in practice, wheelbase often does not have the protective or crash-reducing effect anticipated in the theoretical discussion (i.e., a negative coefficient, denoting longer wheelbase ↔ fewer fatalities). Wheelbase had a negative coefficient in 17 of the 27 basic regressions and positive in 10, not significantly different from a 50-50 split, including 5 negatives in the 9 car regressions and 5 in the 9 LTV regressions. Thus, for cars and LTVs, wheelbase did not have a protective effect in nearly half the regressions. Wheelbase had a significant negative coefficient in only two of the 27 regressions (LTV-fixed object and CUV-other crashes), but a significant positive coefficient in three (car-heavy LTV, LTV-pedestrian, LTV-heavy LTV). In practice, wheelbase appears not so much as to clarify relationships between mass, size, and fatality risk as to muddle them by adding yet another closely related independent variable. The fact that VIF does not increase is perhaps also not that meaningful in logistic regression, for Section 4.5 will show that VIF can be substantially reduced by indexing footprint to mass or vice-versa, yet the logistic regression coefficients remain unchanged.

Wenzel's updated regression analyses combining all crash types likewise find effects of track width and wheelbase that are inconsistent across vehicle types. For passenger cars, the 0.417 inch reduction of track width that historically accompanies a 100-pound reduction of mass is associated with a 1.82 percent fatality increase, a stronger effect than the 1.38 percent increase for a corresponding reduction of footprint (0.737 square feet), whereas the corresponding reduction of wheelbase (1.116 inches) is associated with a 0.10 percent reduction in fatalities. This result could be interpreted as: (1) track width has a genuinely stronger relationship to fatality risk than footprint; or (2) there are too many variables, and track width gets a strong effect in one direction merely to offset the effect for wheelbase in the opposite direction. But for CUVs and minivans, it is track width that has little effect (0.04% fatality increase) while a corresponding reduction of wheelbase strongly increases fatalities (by 1.86%, versus 1.70% for footprint reduction). In truck-based LTVs, track width has a modest effect in the wider ↔ safer direction (0.50%), while wheelbase has a small effect in the opposite direction.[133]

Table 4-1 shows that regression with track width and wheelbase (test no. 3) impacts the Volpe coefficient for light cars and light LTVs by 0.59 percentage points in the direction of less harm/more benefit for mass reduction, with about half as much impact on the heavier cars and LTVs and a slight effect in the opposite direction for CUVs and minivans.

---

[133] Wenzel (2012), Table ES-3, p. x estimates overall effect on fatalities for reducing track width or wheelbase by 1 inch or footprint by 1 square foot. Van Auken and Zellner (2012d), pp. 13 and 17 find the average reductions in track width (inches), wheelbase (inches), and footprint (square feet) as vehicles become 100 pounds lighter, namely: .417, 1.116, and .737 for cars; .470, 1.812, and 1.121 for truck-based LTVs; and .494, 1.603, and .987 for CUVs and minivans.

Without CY control variables:

	Cars < 3,106 lbs	Cars ≥ 3,106 lbs	CUVs & Minivans	LTVs < 4,594 lbs	LTVs ≥ 4,594 lbs
Baseline	1.56	.51	- .37	.52	- .34
W/O CY control variables	1.53	.43	.04	1.20	.30

Alert to opportunities to reduce the number of independent variables in the models, Wenzel sensitivity-tested and discussed a model without the six CY control variables.[134] There is little to add here, except to perform the same analysis with the latest data. To review: two conditions are necessary for the CY variables to have a statistically meaningful impact:

- The mass and or size of the vehicles must change over time – e.g., larger and heavier in later model years.
- Fatality risk must change over time, for example, due to:
    - General improvement in road safety
    - A change in the fleet of potential partner vehicle for crashes, such as fewer light cars as potential partners because light cars are being retired
    - A gradual secular decline in annual VMT per vehicle that is, however, not reflected in the unchanging annual-VMT estimates (Table 2-3).

Both conditions are present: as discussed in Section 1.4, all types of vehicles gained mass and footprint from MY 2000 to 2007, ranging from an average increase of 157 pounds and 1.2 square feet in cars to an increase of 825 pounds and 5.4 square feet in pickup trucks. Fatality rates per VMT generally dropped from CY 2002 to 2008 due to a variety of driver, environmental, and vehicle factors not explicitly included as control variables in the model; they dropped especially for collisions with light cars and light LTVs as there were fewer of them on the road to hit.

Because cars only became a little heavier, controlling for CY might not have much effect in the car regressions. But because LTVs and CUVs grew a lot, controlling for CY could be important. Without it, the regressions would look at the low fatality risk in the later CY and say, "That must be due to the increased mass of the vehicles, because there does not seem to be any other explanation." That is exactly what happened. Table 4-1 (test no. 4) shows that the Volpe coefficients for cars changed little, but the negative coefficients for heavy LTVs and CUVs changed to positive, and the estimated harm for mass reduction in light LTVs came close to the effect in light cars. The results are interesting as a sensitivity test, but cannot be considered accurate for LTVs, because the model erroneously attributes the long-term safety improvement of these vehicles to increased mass.

CUVs and minivans weighted according to 2010 sales:

	Cars < 3,106 lbs	Cars ≥ 3,106 lbs	CUVs & Minivans	LTVs < 4,594 lbs	LTVs ≥ 4,594 lbs
Baseline	1.56	.51	- .37	.52	- .34
CUVs/minivans wtd 2010 sales			.53		

---

[134] Wenzel (2011), pp. 39-46.

In his peer review, Green asks, "Does it make any difference when minivans are deleted from the CUV and minivan group? Compared to the other vehicle types, minivans likely represent a small percentage of vehicles, but also may be quite different from CUVs."[135] Minivans represent a small percentage of vehicles in recent model years. The market share of minivans steadily decreased after MY 2000, while the share for CUVs greatly increased. For example, Table 1-4 shows that minivans accounted for 8.60 percent of sales in MY 2000 and just 4.60 percent in 2007, while the share for CUVs increased from 1.59 to 15.74 percent. In NHTSA's MY 2000-2007 database, the ratio of CUV to minivan registration years is 55.99 to 44.01, but this is no longer representative of sales in a more recent year, say 2010: 85.28 percent CUVs to 14.72 percent minivans. The MY 2000-2007 database is adjusted to mimic MY 2010 sales by multiplying the original weight factor for each CUV case (fatal and induced-exposure) by 85.28/55.99, while multiplying the weight factor for each minivan case by 14.72/44.01.

Table 4-1 (test no. 5) shows that the Volpe coefficient for CUVs and minivans changed from a 0.37 percent fatality reduction to a 0.53 percent increase. The other Volpe coefficients are unchanged, because the baseline analysis carries over for cars and truck-based LTVs. For the individual crash types, giving lower weight to the minivan cases changes the baseline 3.61 percent benefit for mass reduction in fixed-object impacts to a modest 1.64 percent. In three types of crashes (hit car < 3,082, hit LTV ≥ 4,150, all others) where the baseline analysis estimated a small benefit for mass reduction, the new analysis estimates a small fatality increase. In general, this "mostly-CUV" analysis produces results similar to the baseline regressions for heavier cars (except in rollovers, where mass has a strong negative coefficient and collisions with LTVs < 4,150, strong positive – both possibly symptoms of multicollinearity, as in the baseline analysis of CUVs and minivans). (Basing the analysis on CUVs only – simply deleting all the minivan cases rather than re-weighting them – escalates the Volpe coefficient just a bit more, to a 0.78% fatality increase.)

This sensitivity test is useful for illustrating the fragility of the CUV/minivan analysis and, in particular, suggests the benefit for mass reduction estimated in the baseline analysis is a soft number. However, NHTSA does not choose it to supersede the baseline analysis, because of the added complexity of re-weighting the data.

Without non-significant control variables:

	Cars < 3,106 lbs	Cars ≥ 3,106 lbs	CUVs & Minivans	LTVs < 4,594 lbs	LTVs ≥ 4,594 lbs
Baseline	1.56	.51	- .37	.52	- .34
W/O non-sig control vars	1.64	.68	- .46	.35	- .54

In his peer review, Farmer worries that the baseline model may be "overspecified" by having too many covariates – i.e., control variables, namely independent variables other than curb weight and footprint. He recommends examining the sensitivity of the model to deleting some control variables.[136] A possible approach is to delete non-significant (p > .05) control variables one-by-one by backward selection. In other words, for each of the 27 basic regression analyses, start

---
[135] Docket No. NHTSA-2010-0152-0037, p. 12.
[136] Docket No. NHTSA-2010-0152-0036.

with the baseline model and allow the LOGISTIC procedure in SAS® to identify the control variable with the smallest Wald chi-square, run a new regression excluding that variable, and repeat the procedure until all control variables have Wald chi-square ≥ 3.84. Throughout the procedure, the curb-weight and footprint variables are not candidates for deletion and stay in every regression.

The nine baseline regressions for passenger cars usually include 27 control variables (but only 25 for rollovers and 24 for pedestrian crashes, as some or all side-air-bag variables are dropped). The procedure deletes as few as 4 or as many as 11 of these variables from the various regressions. The LTV regressions all begin with 27 control variables; between 6 and 14 drop out. The CUV/minivan regressions begin with 25 to 28 control variables and delete 8 to 15. In many cases, the procedure deletes some of the six CY variables or the eight driver-age variables, while retaining others.

Table 4-1 (test no. 6) shows that deleting non-significant control variables has negligible impact on the Volpe coefficients. None changes by more than 0.20 percentage points. The car coefficients become a trace more positive (more harm for mass reduction) while the LTV and CUV/minivan coefficients show a trace more benefit for mass reduction. Coefficients for individual regressions also stay about the same, although sometimes changing by as much as a percentage point.

This test helps to show that logistic regression is not perturbed by large numbers of basically orthogonal control variables. If they are of little importance, the regression essentially ignores them by giving them small coefficients, hardly changing the coefficients for the other variables. What logistic regression sometimes dislikes is the inclusion of two or more nearly collinear variables such as curb weight, footprint, track width, or wheelbase; it hardly ever cares about the other control variables. An advantage of deleting independent variables is that it shrinks standard errors of the regression coefficients for mass and footprint, usually by less than 10 percent, but considerably more in a few regressions, up to half the error.

This is a good analysis, but, in a sense, it fixes something that is not broken. The baseline analysis has the advantage of applying the same, uniform set of control variables in each regression.

Including muscle/police/AWD cars and full-size vans:

	Cars < 3,106 lbs	Cars ≥ 3,106 lbs	CUVs & Minivans	LTVs < 4,594 lbs	LTVs ≥ 4,594 lbs
Baseline	1.56	.51	- .37	.52	- .34
W. muscle/police/AWD/big van	1.81	.49		.49	- .76

Wenzel tested a model whose regressions include every vehicle in the database, rather than excluding certain niche vehicles, as in the baseline model.[137] Those vehicles are muscle, police, and AWD cars and full-size vans. The CUV/minivan regressions stay the same. The test is already discussed in Section 3.5 of this report; Table 3-7 shows the detailed results. Categorical

---
[137] Wenzel (2011), p. 48.

variables are added to the car and LTV regressions to denote the niche vehicles: police car and muscle car in the car regressions, as well as the AWD variable; cargo van and passenger van in the LTV regressions.

Table 4-1 (test no. 7) shows small-to-moderate changes from the baseline. The Volpe coefficient for cars < 3,106 pounds strengthened by 0.25 percentage points, from a 1.56 percent fatality increase in the baseline to 1.81 here. But the coefficient for LTVs ≥ 4,594 pounds became 0.42 percentage points more negative, from -.34 to -.76.

This alternative model, with its plus of using every vehicle case in the databases, may be one of the most viable substitutes for the baseline model. Nevertheless, NHTSA believes including niche vehicles, each with its own pattern of crash types and of relationships with mass and footprint may distort how the regression allocates effects between mass and footprint. It might generate coefficients for mass and footprint that to some extent reflect how the vehicle mix varies for different mass-footprint combinations rather than the underlying relationships of mass and footprint with fatality risk. There is a trade-off between inclusiveness and uniformity in the data. NHTSA's 2010 report, for example, considered various subsets of cars and found that economy- and sporty 2-door cars could be included with 4-door cars without much change in the coefficients for curb weight and footprint (or curb weight, track width, and wheelbase), but that including muscle cars substantially altered the coefficients.[138]

Control for vehicle manufacturer and control for manufacturer/nameplate:

	Cars < 3,106 lbs	Cars ≥ 3,106 lbs	CUVs & Minivans	LTVs < 4,594 lbs	LTVs ≥ 4,594 lbs
Baseline	1.56	.51	- .37	.52	- .34
Control for veh manuf	1.91	.75	1.64	.68	- .13
for manuf/nameplate	2.07	1.82	1.31	.66	- .13

Wenzel tested a model with NHTSA's preliminary database that included 18 categorical variables denoting a manufacturer and/or nameplate, plus the other baseline control variables.[139] Chen and Kockelman, in their review of Wenzel's 2011 report, recommend splitting this into two analyses.[140] The first adds 15 variables denoting manufacturers only – e.g., Ford, Toyota…. The second adds a total of 20 variables by treating five luxury nameplates as if they were separate manufacturers – e.g., Ford (which now means Ford excluding Lincoln), Lincoln, Toyota (which now means Toyota excluding Lexus), Lexus…. Both techniques are applied here to NHTSA's updated database.

Table 4-1 (test no. 8) shows controlling for manufacturer impacts all five Volpe coefficients in the direction of more harm/less benefit for mass reduction. The impact is great in CUVs and minivans, changing the Volpe coefficient from -0.37 to +1.64. For the other four vehicle types, the impact is modest, ranging from 0.14 to 0.35 percentage points. For CUVs and minivans, controlling for manufacturer changes the effect of mass reduction in fixed-object crashes from a

---

[138] Kahane (2010), p. 482-490.
[139] Wenzel (2011), pp. 38-39.
[140] Menard (2012).

strong to a slight benefit and it changes the effect of mass reduction from a benefit to harm in three other crash types. The design of CUVs and minivans is not that different from passenger cars; with this test, the estimated effects of mass reduction also become fairly similar.

Table 4-1 (test no. 9) shows that controlling for luxury nameplate in addition to manufacturer has a large impact on cars ≥ 3,106 pounds. A high proportion of the vehicles with the five luxury nameplates are cars ≥ 3,106 pounds. The Volpe coefficient for these cars is a 1.82 percent fatality increase per 100-pound mass reduction, almost as high as for the cars < 3,106 (2.07%). It is an escalation of 1.31 percentage points from the baseline and 1.07 percentage points from the previous test controlling only for manufacturer. After controlling for manufacturer and nameplate, the effect of mass reduction in rollovers and in fixed-object impacts becomes harmful in both the heavier and the lighter cars, whereas it was beneficial in both in the baseline analysis. Furthermore, in collisions with heavy trucks and with the heavier LTVs, the effects of mass reduction in heavier cars has become almost as harmful as in lighter cars, whereas in the baseline model, the effects in heavier cars is considerably smaller and non-significant.

In their review of Wenzel's 2011 reports, Chen and Kockelman comment that "the type of car is very much a proxy for driver type…Simply including gender and age variables cannot account for important covariates such as education, risk aversion, driving ability, wealth, etc…These variables…are not readily available in data sets."[141]

In fact, Wenzel's strategy of controlling for manufacturer and/or nameplate is an excellent step toward finding a "proxy for driver type," to the extent that each nameplate has its own brand image and customer following. At the same time, it controls to some extent for differences in vehicle design, at least those differences at the manufacturer level. Furthermore, in the baseline regression analyses without the manufacturer variables, footprint may be acting to some extent as a surrogate for manufacturer – i.e., some nameplates attract good drivers and also tend to have large footprint; without knowing the nameplate, the regression attributes the low fatality risk to the large footprint. Controlling for manufacturer avoids that.

NHTSA believes that controlling for manufacturer and/or nameplate makes sense intuitively. It tones down or eliminates some perhaps puzzling results observed in the baseline analysis, such as the small effects in heavier cars' collisions with heavy trucks or large LTVs, the disparity of the results for CUVs and larger cars, and the benefit for mass reduction in fixed-object impacts. Nevertheless, the agency is unwilling to make it the primary analysis, because it adds so many variables. Also, it is not so clear whether the baseline analysis erroneously attributed effects of manufacturer to footprint or, on the contrary, the new analysis might be erroneously attributing effects of footprint to manufacturer.

Conceptually, even better control for vehicle design could be achieved by adding a categorical variable for every make and model, not just for every manufacturer. NHTSA has performed logistic regressions controlling for make and model when the database was limited to a small

---

[141] Menard (2012).

number of makes and models – e.g., 15 models in an evaluation of side impact protection.[142] But here, it could add hundreds of variables to the analysis and result in an overspecified model.

Wenzel recently added two sensitivity tests, each adding one control variable to the baseline analysis. The first one adds the vehicle's original purchase price; the second adds an estimate of the median owner income, aggregated at the make-model level.[143] The tests were added partly in response to Chen and Kockelman's peer-review recommendation to consider additional driver characteristics.

From Wenzel (2012)[144]	*Cars < 3,106 lbs*	*Cars ≥ 3,106 lbs*	*CUVs & Minivans*	*LTVs < 4,594 lbs*	*LTVs ≥ 4,594 lbs*
Baseline	1.55	.51	- .38	.52	- .34
Control for vehicle price	1.42	.84	- .92	.45	- .52
Control for owner income	1.20	.16	- .44	.68	- .30

Unlike the analyses controlling for vehicle manufacturer, these tests diminished the Volpe coefficient by .13 and .35 percentage points, respectively, relative to baseline, in cars < 3,106 pounds. They made mass reduction more beneficial in CUVs and minivans (unlike the tests controlling for manufacturer, which changed the negative Volpe coefficient to a positive). The impacts on the coefficients for cars ≥ 3,106 pounds and for both groups of truck-based LTVs were relatively small and inconsistent in direction. These two new tests, in combination with the tests controlling for manufacturer or nameplate, show that controlling for various driver and/or vehicle characteristics can push the Volpe coefficients in either direction, depending on the vehicle type and the control variable.

Limited to drivers with BAC=0:

	Cars < 3,106 lbs	Cars ≥ 3,106 lbs	CUVs & Minivans	LTVs < 4,594 lbs	LTVs ≥ 4,594 lbs
Baseline	1.56	.51	- .37	.52	- .34
Limited to BAC=0	2.32	1.06	- .19	.86	- .58

In his peer review of NHTSA's 2010 report, Green asks, "Would deleting alcohol-related crashes from the FARS data change any of the results?"[145] NHTSA's 2011 preliminary report already includes analyses in which the fatal crash cases are limited to case-vehicle drivers with tested or imputed BAC < .01. Table 3-14 of this report updates that analysis with the current databases. Specifically, crash involvements on FARS where the median of the driver's ten actual or imputed values of BAC was .01 or higher were deleted. No cases were deleted from the induced-exposure file, where BAC would usually be unreported for these mostly nonfatal

---

[142] Kahane, C. J. (2007). *An Evaluation of Side Impact Protection – FMVSS 214 TTI(d) Improvements and Side Air Bags*, NHTSA Technical Report No. DOT HS 810 748. Washington, DC: National Highway Traffic Safety Administration, http://www-nrd.nhtsa.dot.gov/Pubs/810748.PDF, pp. 57-59.

[143] Wenzel (2012), pp. 59, 68-69, and 77-79; Wenzel used a database of California vehicle registrations from 2010 to estimate the average income of the household owning the vehicle, based on the zip code of its registered owner; he used the median household income for each zip code in California from the 2000 US Census.

[144] Wenzel (2012), Table ES.2, p. ix.

[145] Docket No. NHTSA-2010-0152-0022, p. 32.

crashes. However, even the FARS data suggests such deletions would have been few. Whereas 21.1 percent of all drivers of MY 2000-2007 cars and LTVs on CY 2002-2008 FARS had reported or imputed BAC ≥ .01, only 4.4 percent of the non-culpable drivers in multi-vehicle crashes had BAC ≥ .01, and only 1.6 percent of the non-culpable drivers of the vehicles with VEH_NO ≥ 3 (who are often bystanders; the principal collision typically is vehicle no. 1 with vehicle no. 2, and one of them may then ricochet into vehicle no. 3). In nonfatal non-culpable induced-exposure involvements, the proportion of drivers with BAC ≥ .01 may be similarly low.

As shown in Table 3-14 and discussed in Section 3.10, the Volpe coefficient is computed by weighting each of the nine crash types by the number of fatalities involving case-vehicle drivers with BAC < .01 (adjusted for ESC). About half the rollovers and fixed-object impacts involved drivers with BAC ≥ .01, versus only about 10 percent of the drivers in the other crash types. In other words, rollovers and impacts with fixed objects, where mass reduction is beneficial for cars and CUVs/minivans, make a smaller contribution to the Volpe coefficient than in the baseline analysis. That pushes the Volpe coefficients for cars and CUVs/minivans in the direction of more harm/less benefit for mass reduction. Table 4-1 (test no. 10) shows a fatality increase of 2.32 percent for cars < 3,106 pounds and 1.06 percent for cars ≥ 3,106 pounds (versus 1.56% and 0.51% in the baseline analysis).

Limiting to sober drivers also tends to intensify the regression coefficients for the collisions involving two light vehicles: make the positives more positive and the negatives more negative. Section 3.10 suggests that limiting the data to sober drivers may have eliminated many of the crashes preceded by loss of control. Having a lighter vehicle might have helped the driver keep control of it and avoid the crash, but when two vehicles slam into one another without prior loss of control, momentum considerations are paramount: net harm for removing mass from the lighter vehicle, net benefit from the heavier vehicle. Thus, in the heavier LTVs, limiting to sober drivers pushes the Volpe coefficient in the opposite direction, up to a 0.58 percent societal benefit (from 0.34% in the baseline analysis).

NHTSA would not contemplate making this the primary analysis. To the agency, the estimated effect on "societal" fatality risk means all of society, not just the sober part of it. Nevertheless, it is an interesting sensitivity test because it shows how controlling for one aspect of driver "quality," in this case, sobriety intensifies rather than weakens the Volpe coefficients for passenger cars. That is consistent with the results of the two preceding tests controlling for manufacturer and/or nameplate.

Limited to good drivers:

	Cars < 3,106 lbs	Cars ≥ 3,106 lbs	CUVs & Minivans	LTVs < 4,594 lbs	LTVs ≥ 4,594 lbs
Baseline	1.56	.51	- .37	.52	- .34
Limited to good drivers	3.00	1.62	.00	1.09	- .30

Wenzel tested another model with NHTSA's preliminary database that excluded drinking drivers and, furthermore, several other groups of drivers exhibiting imprudent behavior in the crash or on previous occasions, as defined in NHTSA's 2003 report.[146] Wenzel's analysis is modified here by excluding not only reported but also imputed BAC ≥ .01 – i.e., every drinking driver excluded in the preceding test. Fatal-crash cases were also excluded if the driver exhibited any one (or more) of these symptoms of imprudent driving behavior:

- Drug involvement on this crash (DRUGS = 1)
- Driving without a valid license at the time of this crash (L_STATUS = 0-4)
- This crash involves driving on a suspended/revoked license, reckless/erratic/negligent driving, being pursued by police, racing, hit & run, or vehicular homicide (any of DR_CF1, DR_CF2, DR_CF3 or DR_CF4 = 19,36[147],37,46,90,91)
- 2 or more reported crashes during the past 3 years (PREV_ACC = 2-75)
- 1 or more DWI convictions during the past 3 years (PREV_DWI = 1-75)
- 2 or more speeding convictions during the past 3 years (PREV_SPD = 2-75)
- 2 or more license suspensions or revocations during the past 3 years (PREV_SUS = 2-75)
- 2 or more other harmful moving violations during the past 3 years (PREV_OTH = 2-75)

As in the preceding test, only fatal crash involvements were deleted. No cases were deleted from the induced-exposure file, where BAC and driver history would usually be unreported for these mostly nonfatal crashes. Unlike the preceding test, the FARS data suggests such deletions might not have been so few. Whereas 36.1 percent of all drivers of MY 2000-2007 cars and LTVs on CY 2002-2008 FARS exhibited one or more symptoms of imprudent driving, so did 15.4 percent of the non-culpable drivers in multi-vehicle crashes and even 11.4 percent of the non-culpable drivers of the vehicles with VEH_NO ≥ 3 ("bystanders"). If that 11.4 percent were also characteristic of the nonfatal induced-exposure crashes, although it is not a large percentage overall, it could be substantially higher in some makes and models. A caveat on the analysis is that the results might be different if it had been possible to exclude corresponding cases from the induced-exposure database. Wenzel suggests controlling for driver characteristics that are already available in the State data or from other sources.[148] Another possibility in future research would be to obtain driver-history information for the induced-exposure cases by the same method as FARS analysts obtain if for fatality cases, namely by linking crash data to driver-history files.

Table 4-1 (test no. 11) shows that limiting to good drivers strengthens the Volpe coefficient for cars < 3,106 pounds to 3.00 percent, nearly double the baseline (1.56%). The coefficient for cars ≥ 3,106 pounds rises to 1.62 percent and for LTVs < 4,594 pounds to 1.09 percent (versus 0.51% and 0.52%, baseline). The overall effect becomes zero in CUVs and minivans. Only the Volpe coefficient for heavy LTVs stays about the same (-0.30% rather than -0.34%) and continues to show a benefit for mass reduction.

The Volpe coefficient is computed by weighting each of the nine crash types by the number of fatalities involving good drivers only. That eliminates 62 percent of the rollovers and fixed-object impacts, versus only 26 percent of the drivers in the other crash types. In other words,

---

[146] Wenzel (2011), pp. 46-47; Kahane (2003), p. 94.
[147] In Florida, Kansas, North Carolina, Ohio, and Utah, do not include if DR_CF2, DR_CF3 or DR_CF4 = 36, since that code is applied frequently in those States and does not necessarily mean reckless driving.
[148] Wenzel (2012), pp. 78-79.

rollovers and impacts with fixed objects, where mass reduction is most beneficial for cars and CUVs/minivans, make a considerably smaller contribution to the Volpe coefficient than in the baseline analysis. But furthermore, for these good drivers of passenger cars, mass reduction is less beneficial or not beneficial at all in rollovers and fixed-object impacts, and more harmful than in the baseline analysis in the other collision types. It is the same pattern as in the preceding test, but stronger. This test, even more than the three preceding ones, shows how controlling for driver "quality" can intensify the Volpe coefficients for passenger cars.

Perhaps a take-home message from the last two sensitivity tests is that impaired or incautious drivers, on the relatively frequent occasions when they find their vehicles out of control or pointed in the wrong direction, initiate emergency maneuvers; under those circumstances, any extra mass makes it even more difficult for them to regain directional control. A vehicle combining small footprint and high mass may be toxic for these drivers. By contrast, good, cautious drivers rarely need such maneuvers. For them, the paramount effect of mass is conservation of momentum. For them, mass reduction is especially harmful in the lightest cars, while continuing to be societally beneficial in the heavier LTVs – exactly as shown in the two sensitivity tests.

4.4   Effect on the case vehicle's versus the other vehicle's occupants

In his peer review, Farmer recommends, "Although the purpose of the analysis was to examine the effects of weight reduction on societal risk, it should be shown that weight reductions [in the case vehicle] increase the risk to occupants of the case vehicle (as expected from the conservation of momentum). This makes for another validation of the methodology."[149] By the same token, mass reduction in the case vehicle ought to mitigate the risk to occupants of the other vehicle, in a collision between two light vehicles. Changing the footprint of the case vehicle, on the other hand, should not show as strongly contrasting effects between the case and the other vehicles, because conservation of momentum is not an issue. There could be one effect primarily in the case vehicle, namely if extra footprint provides more crush space for the occupants of the case vehicle. There could also be another effect in the same direction for both vehicles, namely if extra footprint helps maintain stability and prevent the crash entirely (although not that many multi-vehicle collisions are preceded by loss of control and there could be an offsetting effect that a longer and wider vehicle is a larger target while under control).

All of these hypotheses can be investigated by simply performing the baseline regressions for the four types of crashes that involve hitting another light vehicle (collision with car, CUV, or minivan < 3,082 pounds; collision with car, CUV, or minivan ≥ 3,082 pounds; collision with truck-based LTV < 4,150 pounds; and collision with truck-based LTV ≥ 4,150 pounds), except replacing the baseline dependent variable – fatalities in the crash – with the number of occupant fatalities in the <u>case</u> vehicle. Then perform the regressions again with the dependent variable being the occupant fatalities in the <u>other</u> vehicle. Table 4-4 shows the regression coefficients for mass in the 12 regressions and their Wald chi-squares. The left columns just recapitulate the baseline regressions from Table 3-1. The middle columns are the effect on fatality risk in the case vehicle; the right columns, the risk in the other vehicle. Table 4-5 presents the corresponding regression coefficients for footprint. As in the baseline analysis, there are

---

[149] Docket No. NHTSA-2010-0152-0036.

separate mass coefficients for cars lighter and heavier than 3,106 pounds and LTVs lighter and heavier than 4,594 pounds, but not for CUVs and minivans. There are no separate footprint coefficients for lighter and heavier vehicles of the same type. Black numerals indicate a fatality increase given mass reduction in the case vehicle; red numerals, a fatality reduction. Bold numerals indicate a regression coefficient with Wald chi-square ≥ 3.84.

The left columns of Table 4-4 show that mass reduction in the lighter vehicles – passenger cars – increases societal fatality risk in 8 of 8 regressions, while mass reduction in the heavier vehicles – LTVs – has societal benefits in 6 of 8 regressions, the two exceptions being when they hit somebody their own mass or heavier. Five of the 20 coefficients have Wald chi-square ≥ 3.84, three in the direction of mass reduction $\leftrightarrow$ societal harm and two in the opposite direction.

The middle columns of Table 4-4 show mass reduction in the case vehicle is almost always harmful for occupants of the case vehicle in the collisions of two light vehicles: 19 of the 20 coefficients are in the direction of mass reduction $\leftrightarrow$ harm to case vehicle occupants and 12 of these 19 have Wald chi-square ≥ 3.84. The only red number (non-significant) is when LTVs $\geq$ 4,594 pounds hit cars $< 3,082$ pounds, a low-risk situation for the occupants of the LTVs. The average of the 20 coefficients is 4.32. While simply averaging the coefficients is not rigorous, it at least suggests that mass has a strong protective effect for occupants of the case vehicle.

The right columns of Table 4-4 demonstrate that mass reduction in the case vehicle is almost always beneficial for occupants of the other vehicle: 19 of the 20 coefficients are in the direction of case-vehicle mass reduction $\leftrightarrow$ benefit to other-vehicle occupants and 8 of these 19 have Wald chi-square ≥ 3.84. The exception (non-significant) is when light LTVs hit heavy LTVs, a low-risk situation for the occupants of the heavy LTVs. The average of the 20 coefficients is -2.38. The results are what would be expected based on conservation of momentum and they would appear to be the "validation of the methodology" that Farmer recommended.

The coefficients for footprint reduction in Table 4-5 do not show the same sort of contrast between the case vehicle and the other vehicle. The left columns indicate the societal effect of footprint reduction. It is usually small and more often than not in the direction of footprint reduction $\leftrightarrow$ societal harm. However, the results are muddied by several large coefficients, both positive and negative, that perhaps indicate symptoms of multicollinearity rather than a true effect.

The middle columns of Table 4-5 show a harmful effect to the case-vehicle occupants for footprint reduction in 9 of 12 regressions. But Wald chi-square ≥ 3.84 in only two of these nine, and it also exceeds 3.84 for one of the negative coefficients. Furthermore, two of the three significant coefficients (cars hitting LTVs $< 4,150$ pounds and CUVs/minivans hitting LTVs $< 4,150$ pounds) may be suspected of symptoms of multicollinearity because of similarly large coefficients (one positive, one negative) in the societal analysis. The 12 coefficients in the middle columns of Table 4-5 average out to 0.53; excluding the two suspect coefficients (+5.81 and -9.32), the remaining ten average out to 1.00. While simply averaging the coefficients is not rigorous, it at least suggests that footprint has a modest protective effect for occupants of the case vehicle, but far less than the corresponding average effect of mass (4.32).

TABLE 4-4: EFFECTS OF 100-LB MASS REDUCTION ON CRASH, CASE-VEHICLE AND OTHER-VEHICLE FATALITIES BY CASE-VEHICLE CLASS AND CRASH TYPE

CRASH TYPE	CRASH FATALITIES		CASE-VEHICLE FATALITIES		OTHER-VEHICLE FATALITIES	
	FATALITY INCREASE (%)	WALD CHI-SQUARE	FATALITY INCREASE (%)	WALD CHI-SQUARE	FATALITY INCREASE (%)	WALD CHI-SQUARE
CARS < 3,106 POUNDS						
HIT CAR/CUV/MINIVAN < 3082	.76	.57	6.95	16.72	- 3.21	6.44
HIT CAR/CUV/MINIVAN 3082+	.48	.25	5.37	18.48	- 7.34	24.26
HIT TRUCK-BASED LTV < 4150	1.17	.98	2.23	2.44	- 1.76	.67
HIT TRUCK-BASED LTV 4150+	6.06	29.80	7.29	36.47	- 1.43	.25
CARS ≥ 3,106 POUNDS						
HIT CAR/CUV/MINIVAN < 3082	.26	.05	2.57	1.52	- 1.01	.48
HIT CAR/CUV/MINIVAN 3082+	1.62	2.09	5.14	11.53	- 2.78	2.79
HIT TRUCK-BASED LTV < 4150	.53	.14	.64	.14	- .41	.03
HIT TRUCK-BASED LTV 4150+	2.34	3.11	3.86	6.99	- 5.42	2.84
PICKUPS & TRUCK-BASED SUVs < 4,594 POUNDS						
HIT CAR/CUV/MINIVAN < 3082	- .09	.02	4.63	9.01	- 1.25	3.90
HIT CAR/CUV/MINIVAN 3082+	- .71	1.30	2.34	3.26	- 2.50	12.07
HIT TRUCK-BASED LTV < 4150	- .63	.56	4.08	7.98	- 4.21	15.38
HIT TRUCK-BASED LTV 4150+	4.46	26.58	5.76	25.24	.83	.37
PICKUPS & TRUCK-BASED SUVs ≥ 4,594 POUNDS						
HIT CAR/CUV/MINIVAN < 3082	- .91	3.99	.41	.08	.77	2.53
HIT CAR/CUV/MINIVAN 3082+	- 1.37	8.05	.42	.11	- 1.32	6.15
HIT TRUCK-BASED LTV < 4150	- .96	1.99	5.50	11.10	- 2.13	7.87
HIT TRUCK-BASED LTV 4150+	.53	.53	4.01	11.59	- .95	.91
CUVs & MINIVANS						
HIT CAR/CUV/MINIVAN < 3082	- .09	.01	3.04	.79	- .55	.15
HIT CAR/CUV/MINIVAN 3082+	1.68	1.54	10.67	17.91	- 2.30	2.05
HIT TRUCK-BASED LTV < 4150	3.82	3.94	11.06	12.96	- 1.23	.24
HIT TRUCK-BASED LTV 4150+	- .93	.29	1.22	.37	- 7.83	5.21

TABLE 4-5

EFFECTS OF 1 SQ FT FOOTPRINT REDUCTION ON CRASH, CASE-VEHICLE AND OTHER-VEHICLE FATALITIES BY CASE-VEHICLE CLASS AND CRASH TYPE

CRASH TYPE	CRASH FATALITIES		CASE-VEHICLE FATALITIES		OTHER-VEHICLE FATALITIES	
	FATALITY INCREASE (%)	WALD CHI-SQUARE	FATALITY INCREASE (%)	WALD CHI-SQUARE	FATALITY INCREASE (%)	WALD CHI-SQUARE
			CARS			
HIT CAR/CUV/MINIVAN < 3082	.23	.04	2.28	1.32	- 1.30	.85
HIT CAR/CUV/MINIVAN 3082+	.49	.20	.72	.24	- .52	.10
HIT TRUCK-BASED LTV < 4150	3.96	8.30	5.81	12.22	- .55	.05
HIT TRUCK-BASED LTV 4150+	1.77	1.89	1.45	1.06	2.52	.63
			PICKUPS & TRUCK-BASED SUVs			
HIT CAR/CUV/MINIVAN < 3082	- .21	.24	1.18	.96	- .38	.70
HIT CAR/CUV/MINIVAN 3082+	.31	.46	2.65	6.84	- .19	.15
HIT TRUCK-BASED LTV < 4150	1.01	2.81	1.52	1.85	1.01	2.00
HIT TRUCK-BASED LTV 4150+	-1.59	0.88	-1.23	1.99	- 2.07	4.74
			CUVs & MINIVANS			
HIT CAR/CUV/MINIVAN < 3082	- .79	.19	3.91	.80	- 1.85	.87
HIT CAR/CUV/MINIVAN 3082+	- 2.19	1.47	- 5.36	2.92	- .92	.17
HIT TRUCK-BASED LTV < 4150	- 4.05	2.48	- 9.32	5.87	- .05	.0002
HIT TRUCK-BASED LTV 4150+	3.80	2.64	2.83	1.15	6.96	2.08

The right columns of Table 4-5 show a benefit to the other-vehicle occupants for footprint reduction in the case vehicle in 9 of 12 regressions. But only one of these nine has Wald chi-square ≥ 3.84 and several of the effects are close to zero. The 12 coefficients average out to 0.22; excluding two possible outliers (+6.96 and -2.07), the remaining ten average out to -0.22. Unlike mass reduction, footprint reduction in the case vehicle does not have a statistically meaningful effect on occupants of the other vehicle.

These additional estimates requested by Farmer supplement the analysis of downsizing (Section 3.7) to remind us cogently that, even though the _societal_ effect of mass reduction while controlling for footprint, if any, is small, the effect of mass reduction on the fatality risk of some people in some types of crashes can be quite substantial in either direction.

4.5 Indexing curb weight to footprint and other ways to address multicollinearity

NHTSA's 2010 report discussed multicollinearity issues at length. In their reviews of the 2010 report, Farmer and Green both recommended finding better ways to lessen the effects of multicollinearity or to address it analytically. One technique they suggested was indexing mass to footprint.[150] NHTSA's 2011 preliminary report responded tersely, "[A] method to index curb weight to footprint [was] considered in response to comments by the peer reviewers. A linear regression estimated the expected curb weight given a vehicle's footprint and body type. The excess weight was defined as the actual minus the expected curb weight…The indexing technique successfully lower[ed] VIF but [did] not affect the main analysis results."[151] The 2011 report (Section 3.4) also asserted that multicollinearity had become less of a problem with the MY 2000-2007 database because, even though perhaps four of the 27 basic regressions exhibited possible symptoms of it, two of them gave a strong positive coefficient to mass and negative to footprint, whereas the other two did the reverse – essentially canceling each other out, at least in the aggregate. In his review of the 2011 report, Farmer again raises the issue and requests more detail on how NHTSA tried to index mass to footprint and what happened.

A good starting point is the baseline regression for first-event rollovers of CUVs and minivans, which appears, at least on paper, to have the most symptoms of multicollinearity among the 27 basic regressions. Here are some of the coefficients:

```
                                Standard       Wald
Parameter      DF    Estimate    Error     Chi-Square    Pr > ChiSq

Intercept       1    -21.8533    0.7359     881.9362      <.0001
LBS100          1      0.0702    0.0179      15.3091      <.0001
FOOTPRNT        1     -0.1159    0.0243      22.6596      <.0001
MINIVAN         1     -0.0101    0.1497       0.0045       0.9465
ROLLCURT        1     -0.7729    0.2713       8.1150       0.0044
ABS             1     -0.0111    0.1812       0.0038       0.9511
ESC             1     -0.6134    0.1816      11.4033       0.0007
And so on...
```

---

[150] Docket items nos. NHTSA-2010-0152-0005 (Farmer) and NHTSA-2010-0152-0022 (Green).
[151] Kahane (2011), p. 57.

Mass reduction by 100 pounds reduces fatality risk by 7.02 percent, whereas footprint reduction by one square foot increases risk by 11.59 percent. But the decile analysis (Table 3-4) does not really confirm such a strong effect for mass reduction; the results of the baseline regression look like symptoms of multicollinearity.

A linear regression of footprint by curb weight in the CUV/minivan induced-exposure cases, weighted by VMT, finds:

$$\text{Typical footprint} = 18.86 + .007714 * \text{curb weight}$$

"Excess footprint" is defined as:

$$EX\_FTP = \text{actual footprint} - \text{typical footprint}$$

Whereas LBS100 and FOOTPRNT had a correlation of .781 for CUVs and minivans, LBS100 and EX_FTP have a correlation of .000. Whereas the maximum VIF for the independent variables was 8.71 with curb weight, footprint, plus the other control variables, it is only 4.23 when EX_FTP is substituted for FOOTPRNT. Here are the coefficients if the regression of fatality risk in rollovers is run with EX_FTP substituted for FOOTPRNT:

```
                            Standard      Wald
Parameter      DF  Estimate    Error   Chi-Square    Pr > ChiSq

Intercept       1  -24.0391    0.4729   2584.4211      <.0001
LBS100          1   -0.0192    0.0131      2.1689      0.1408
EX_FTP          1   -0.1159    0.0243     22.6596      <.0001
MINIVAN         1   -0.0101    0.1497      0.0045      0.9465
ROLLCURT        1   -0.7729    0.2713      8.1150      0.0044
ABS             1   -0.0111    0.1812      0.0038      0.9511
ESC             1   -0.6134    0.1816     11.4033      0.0007
And so on...
```

The coefficients for MINIVAN, ROLLCURT, ABS, ESC, and all the remaining control variables stay exactly the same and so do their chi-squares. The coefficient for EX_FTP is identical to FOOTPRNT in the preceding regression. The only change is the intercept and the coefficient for LBS100, which has reversed from +.0702 to -.0192. At first glance, the multicollinearity issue appears "solved." The new regression predicts large harm for footprint reduction and small, non-significant harm for mass reduction. But LBS100 is no longer really curb weight. Conceptually, it is now a combination of curb weight and the footprint that would be typical for that curb weight. Note that:

$$+.0702 - .7714 * .1159 = -.0192$$

where .0702 is the original coefficient for LBS100, .7714 is the coefficient for curb weight (expressed in hundreds of pounds) in the linear regression of footprint by curb weight, .1159 is the original coefficient for FOOTPRNT, and -.0192 is the new coefficient for LBS100. The new variables called "LBS100" and "EX_FTP" are linear combinations of the old variables LBS100 and FOOTPRNT. The new result does not convey any new information but just illustrates a principle: linear re-combination of some of the independent variables essentially changes nothing in a logistic regression. The new regression is saying exactly the same thing as the baseline.

Specifically, the effect of "removing 100 pounds while maintaining footprint" is estimated in the baseline model by reducing LBS100 by 1 and changing nothing else; the estimated effect is a 7.02 percent decrease in rollover fatalities. But in the model with EX_FTP instead of footprint, LBS100 decreases by 1 while EX_FTP simultaneously increases by 0.7714 (because mass reduction while maintaining footprint will result in an increase of "excess footprint" – the lower-mass vehicle would be expected to have smaller footprint, but it does not). The estimated effect is again

$$.0192 - .7714*.1159 = 7.02 \text{ percent decrease}$$

Conversely, a linear regression of curb weight by footprint finds:

$$\text{Typical curb weight} = 16.76 + 79.007*\text{footprint}$$

"Excess curb weight" (in hundreds of pounds) is defined as:

$$\text{EX\_WT00} = .01*(\text{actual curb weight} - \text{typical curb weight})$$

FOOTPRNT and EX_WT00 have a correlation of .000. VIF again drops from 8.71 to 4.23. Here are the regression coefficients when EX_WT00 is substituted for LBS100:

Parameter	DF	Estimate	Standard Error	Wald Chi-Square	Pr > ChiSq
Intercept	1	-21.8416	0.7366	879.3210	<.0001
EX_WT00	1	0.0702	0.0179	15.3091	<.0001
FOOTPRNT	1	-0.0605	0.0167	13.1753	0.0003
MINIVAN	1	-0.0101	0.1497	0.0045	0.9465
ROLLCURT	1	-0.7729	0.2713	8.1150	0.0044
ABS	1	-0.0111	0.1812	0.0038	0.9511
ESC	1	-0.6134	0.1816	11.4033	0.0007

And so on...

The coefficients for the control variables again stay the same. The coefficient for EX_WT00 is identical to LBS100 in the preceding regression. The only change is the intercept and the coefficient for FOOTPRNT, which attenuated from -.1159 to -.0605. Again:

$$-.1159 + .79007*.0702 = -.0605$$

A second conceivable technique for dealing with symptoms of multicollinearity is to redefine the crash types: specifically, the single-vehicle crash types, where the symptoms most often appear. Regressions may be performed on a single category, "run-off-road crashes," which includes all first-event rollovers and impacts with fixed objects plus two types that the baseline analysis classifies in the "all other" group: (1) first-event immersions and (2) crashes where it is difficult to tell if the first truly harmful event was a rollover or a collision with a fixed object (e.g., contacting a guardrail and rolling over), but the vehicle clearly ran off the road before that event.

For CUVs and minivans, the regression coefficients are .0471 for LBS100 and -.0894 for FOOTPRNT – i.e., between the baseline coefficients for rollovers (.0702 and -.1159) and for fixed-object impacts (.0361 and -.0767). The regression takes the average of the two baseline coefficients, so to speak, and continues to display the same symptoms. For passenger cars, the coefficients are .0076 for UNDRWT00, .0192 for OVERWT00, and -.0498 for FOOTPRNT: again, between the baseline coefficients for rollovers (.0183, .0289, and -.0808) and fixed-object impacts (.0046, .0129, and -.0401). (In the case of passenger cars, unlike CUVs and minivans, the coefficients for mass are relatively small and do not, at first glance, suggest symptoms of multicollinearity; in any case, redefining the crash types does not change the sign.).

A third possible technique is to modify the regression algorithm. In the review of the 2011 report, Farmer asks, "Was logistic ridge regression considered as an alternative? As this analysis was based on a very large sample, ridge estimation may not make any difference, but it should at least be mentioned."[152] Ridge regression, also called penalized regression is an analytic technique for addressing multicollinearity. When there are symptoms of multicollinearity – two highly correlated variables with strong coefficients in opposite directions – it repeatedly weakens the coefficients, checking if that improves overall fit (mean square error). Ridge estimation is primarily designed for least-squares linear regression, but several researchers have adapted it to logistic regression, where it may have an additional advantage of not increasing bias.[153] However, NHTSA did not find code available to the public for general applications of logistic ridge regression in SAS® - e.g., in the LOGISTIC procedure or related procedures. Individual researchers have developed algorithms in SAS® IML and macro languages for their own databases.[154] The FIRTH option in the LOGISTIC procedure is somewhat related but not designed to address multicollinearity; it does not affect the results. The literature notes (like Farmer) that ridge regression makes a difference primarily when the ratio of the sample size to the number of independent variables is "small" – e.g., less than 10 or less than 50.[155] The 27 baseline regressions have 27 to 30 independent variables, each. The sample sizes in each regression range from 622 to 10,487 fatal cases and 371,009 to 1,242,083 induced-exposure cases. Thus, even counting only the fatal cases, the ratio of sample size to number of independent variables is normally well over 50, but it is in the low 20s for three of the CUV/minivan regressions. NHTSA will consider obtaining code to perform ridge regressions in future analyses with multicollinearity issues.

While these analytical techniques were not helpful, NHTSA notes that some of the sensitivity tests did reverse or attenuate the mass coefficients in rollovers and/or fixed-object impacts:

---

[152] Docket No. NHTSA-2010-0152-0036.

[153] Gao, S., and Shen, J. (2007). "Asymptotic Properties of a Double Penalized Maximum Likelihood Estimator in Logistic Regression," *Statistics & Probability Letters*, Vol. 77, pp. 925-930; El-Dereny, M., and Rashwan, N. I. (2011). "Solving Multicollinearity Problem Using Ridge Regression Models," *International Journal of Contemporary Mathematical Sciences*, Vol. 6, pp. 585-600, http://www.m-hikari.com/ijcms-2011/9-12-2011/rashwanIJCMS9-12-2011.pdf.

[154] Shen, J., and Gao, S. (2008). "A Solution to Separation and Multicollinearity in Multiple Logistic Regression," *Journal of Data Science*, Vol. 6, pp. 515-531.

[155] Vago, E., and Kemeny, S. (2006). "Logistic Ridge Regression for Clinical Data Analysis (A Case Study)." *Applied Ecology & Environmental Research*, Vol. 4, pp.171-179.

controlling for manufacturer/nameplate reversed all four coefficients in the direction of mass reduction ↔ higher risk in passenger cars; limiting the analysis to CUVs attenuated the mass coefficient to less than one percentage point in fixed-object collisions.

4.6 Response to other comments

Anders Lie's review of NHTSA's 2011 preliminary report notes:

> "[NHTSA's] results indicate [societal] safety changes in levels around 1% [order of magnitude, per 100-pound mass reduction]. The effects are small compared to the safety differences between car models coming from design and engineering. A recent study on the effect of good Euro NCAP scores shows that the difference in modern cars is significant. For fatalities the difference between 2-star and 5-star cars were 68 ± 32 percent.[156] A discussion around the magnitude of benefits from ESC and intelligent seat belt reminders would be valuable as reference points. My recommendation is to add a short note or paragraph illustrating the magnitude of some other vehicle safety developments and in that way putting the magnitude of the effects of mass or footprint into perspective."[157]

NHTSA concurs. Statistical analyses comparing the relative fatality risk in actual crashes between vehicles with good and poor ratings showed 26 to 74 percent lower fatality risk in the better-rated vehicle, depending on the type of crash. These analyses are discussed in more detail in Section 3.8, as background for the regression analyses including IIHS crash-test ratings.

Here are estimates of fatality reduction for some vehicle technologies, based on statistical analyses of crash data:[158]

		Fatality Reduction (%)	
		Cars	LTVs
Electronic stability control	All fatal crashes	23	20
	First-event rollovers	56	74
Seat belts	All fatalities	45	60
Frontal air bags	All fatalities	12	12
	Purely frontal impacts	29	29
Curtain + torso bags	Nearside impacts	24	24

These key safety technologies likewise have effectiveness at least an order of magnitude larger than the likely range of societal effects of reducing mass by 100 pounds.

---

[156] Kullgren, Lie, and Tingvall (2010).
[157] Docket No. NHTSA-2010-0152-0035.
[158] Sivinski (2011); Kahane, C. J. (2004). *Lives Saved by the Federal Motor Vehicle Safety Standards and Other Vehicle Safety Technologies, 1960-2002*, NHTSA Technical Report No. DOT HS 809 833. Washington, DC: National Highway Traffic Safety Administration, http://www-nrd.nhtsa.dot.gov/Pubs/809833.PDF; Kahane, C. J. (2007). *An Evaluation of Side Impact Protection – FMVSS 214 TTI(d) Improvements and Side Air Bags*, NHTSA Technical Report No. DOT HS 810 748. Washington, DC: National Highway Traffic Safety Administration, http://www-nrd.nhtsa.dot.gov/Pubs/810748.PDF.

Van Auken and Zellner's public comments on NHTSA's 2011 report present a series of statistical analyses. First, they reproduce NHTSA's 2011 baseline analysis: 27 regressions of societal fatality risk per VMT by curb weight, footprint, and control variables. Next, they introduce a regression technique that parses each regression coefficient – for each vehicle type, crash type, and independent variable – into two: (1) the effect on societal <u>fatalities</u> per police-reported <u>accident</u> involvement of that crash type, "F/A" and (2) the effect on police-reported <u>accident</u> involvements (of that crash type) per <u>VMT</u>, "A/VMT." They accomplish this by classifying crash involvements in the State data files – not limited to induced-exposure involvements but including culpable, single-vehicle, and numerous-vehicle crash involvements – into the same nine crash types that NHTSA uses with FARS. By selecting a number of crashes for each State and CY proportional to its vehicle registration years, they are able to create regression printouts that have the same independent variables as NHTSA's baseline analysis but that also implicitly control for reporting differences across States.[159]

For passenger cars and truck-based LTVs, overall and for many of the individual crash types, these analyses tend to show that (1) mass reduction lowers F/A, but (2) increases A/VMT.[160]

The analyses appear to be computationally valid. The sum of the F/A and the A/VMT coefficients is usually close to the baseline coefficients in NHTSA's analysis.

However, in most of their tables, Van Auken and Zellner label the column of F/A coefficients as the "effect of mass reduction on **crashworthiness** and crash compatibility" and the A/VMT coefficients as its "effect on **crash avoidance**." In other words, the tables say mass reduction benefits crashworthiness and harms crash avoidance. NHTSA believes these are not accurate characterizations of the coefficients and they lead, in turn, to misunderstandings. Specifically, the ICCT in their public comment argue that the observed benefit to crashworthiness and harm to crash avoidance is counterintuitive and may be evidence of a flaw in the baseline analysis, such as a need for additional or different control variables.[161]

NHTSA believes the metric of fatalities per reported crash (F/A) does not measure just crashworthiness but also certain important aspects of crash avoidance, namely the severity of a crash. In addition, it could be influenced by how often crashes are reported or not reported.

Conceptually, crashworthiness is the likelihood that an occupant will survive, given an impact to a vehicle that in turn results in a particular physical insult to the occupant. It is quite appropriate for the regression analyses to control for driver age and gender, because it is known that, given the same physical insult, a person is more likely to die with each year that he or she gets older. Furthermore, from young adulthood up to middle age, a female is more likely to die from the same physical insult as a male of the same age. Crash-data analyses have shown increases in fatality risk of 2 to 4 percent for each year that a person gets older. Young adult females are 20 to 30 percent more vulnerable than males of the same age; that differences decreases over time

---

[159] Van Auken and Zellner (2012b), Vol. 1, pp. xix, 35-38, and 64-70 and Vol. 3, (Docket No. NHTSA-2010-0152-0032).
[160] *Ibid.*, pp. xix and 66-69.
[161] Docket No. NHTSA-2010-0131-0258, p. 10.

and eventually reverses by late middle age, but averaging across all ages, females are still 5 to 20 percent more vulnerable than males of the same age.[162]

In other words, if these F/A regressions truly modeled crashworthiness, the analyses of the crash types where most fatalities are in the case vehicle (rollover, fixed-object, heavy-truck, and the various types of collisions where the other vehicle is heavier) should have coefficients like -0.03 for M14_30, F14_30, M30_50, and F30_50, each of which measure how many years the driver is younger than 30 or 50, respectively. They should have coefficients like +0.03 for M50_70, F50_70, M70_96, and F70_96, which measure how many years the driver is older than 50 or 70. They should have a coefficient like -0.10 for DRVMALE, because a male is less vulnerable than a female. In crashes where the fatalities are uncommon in the case vehicle (hitting a pedestrian or a much lighter vehicle), the coefficients should all be close to zero, because the age or gender of the driver will not affect how the pedestrian reacts to a physical insult.

Instead, the regressions rather consistently estimate positive or near-zero coefficients for M14_30, F14_30, M30_50, and F30_50 and positive coefficients for DRVMALE. They say F/A decreases as the occupant ages up to age 50 and F/A is lower for females than males.[163]

A more blatant example: on purely crashworthiness considerations, whether it is light or dark outside ought to have little effect on the risk of death from a given physical insult, except perhaps to the extent it affects EMS arrival. But NITE is consistently associated with an extraordinary increase in F/A.

Of course, it is obvious what is going on. These crash data have no measure of crash severity, such as delta v. M14_30, F14_30, M30_50, F30_50, DRVMALE, and NITE all act as surrogates for crash severity. They not only indicate crashworthiness (ability to survive a physical insult) but also, and in some cases primarily, crash avoidance – namely, the ability of age 30-50 drivers, females, and daytime drivers to stay out of situations that lead to fatal crashes, while having their share of fender-benders. Driving at night, on the other hand, is a way to avoid fender-benders characteristic of rush-hour traffic and thereby increases F/A.

Just as many of the control variables in the F/A regressions measure effects of crash avoidance in addition to (and sometimes in place of) crashworthiness, by the same token, there is no particular reason that the coefficients for UNDRWT00, OVERWT00, and FOOTPRNT measure the effects of crashworthiness exclusively and not also crash avoidance. Control variables such as M14_30, F14_30, M30_50, F30_50, DRVMALE, NITE, and also SPDLIM55 and RURAL may account for much of the effect of crash severity on risk per reported crash, but it is unknown exactly how much.

A salient feature of NHTSA's approach, where the numerator is fatalities and the denominator VMT, is to take crash-reporting rates out of the formula for calculating risk. A fatality is a fatality and a mile of travel is a mile of travel – unlike contact events that may or may not be police-reported, depending on the vehicle, the driver, the locality, or the circumstances of the moment. These analyses of F/A and A/VMT appear to be computationally valid, but NHTSA

---

[162] Evans, L. (1991). *Traffic Safety and the Driver*. New York: Van Nostrand Reinhold, pp. 22-28; Kahane is currently analyzing data on the latest vehicles.
[163] Van Auken and Zellner (2012b), Vol. 3, Appendix E, (Docket No. NHTSA-2010-0152-0032).

doubts they truly measure the "effect of mass reduction on crashworthiness" and "effect of mass reduction on crash avoidance."

References

Allison, P.D. (1999), *Logistic Regression Using the SAS System*. Cary, NC: SAS Institute Inc.

Auto Insurance Loss Facts, September 2009, http://www.iihs.org/research/hldi/fact_sheets/CollisionLoss_0909.pdf.

Bean, J.D., Kahane, C. J., Mynatt, M., Rudd, R.W., Rush, C.J., and Wiacek, C. (2009). *Fatalities in Frontal Crashes Despite Seat Belts and Air Bags*, NHTSA Technical Report. DOT HS 811 202. Washington, DC: National Highway Traffic Safety Administration, http://www-nrd.nhtsa.dot.gov/pubs/811102.pdf.

Blodgett, R. J. (1983). *Pedestrian Injuries and the Downsizing of Cars*. Paper No. 830050. Warrendale, PA: Society of Automotive Engineers.

Cerrelli, E. (1973). "Driver Exposure: The Indirect Approach for Obtaining Relative Measures," *Accident Analysis and Prevention*, Vol. 5, pp. 147-156.

El-Dereny, M., and Rashwan, N. I. (2011). "Solving Multicollinearity Problem Using Ridge Regression Models," *International Journal of Contemporary Mathematical Sciences*, Vol. 6, pp. 585-600, http://www.m-hikari.com/ijcms-2011/9-12-2011/rashwanIJCMS9-12-2011.pdf.

Evans, L. (1991). *Traffic Safety and the Driver*. New York: Van Nostrand Reinhold.

Farmer, C. M. (2005). "Relationships of Frontal Offset Crash Test Results to Real-World Driver Fatality Ratings," *Traffic Injury Prevention*, Vol. 6, pp. 31-37.

FRIA (2010). *Final Regulatory Impact Analysis: Corporate Average Fuel Economy for MY 2012-MY 2016 Passenger Cars and Light Trucks*. Washington, DC: National Highway Traffic Safety Administration, http://www.nhtsa.dot.gov/staticfiles/DOT/NHTSA/Rulemaking/Rules/Associated%20Files/CAFE_2012-2016_FRIA_04012010.pdf

Gao, S., and Shen, J. (2007). "Asymptotic Properties of a Double Penalized Maximum Likelihood Estimator in Logistic Regression," *Statistics & Probability Letters*, Vol. 77, pp. 925-930.

Greene, W. H. (1993). *Econometric Analysis*, Second Edition. New York: Macmillan Publishing Company.

Haight, F.A. (1970). "A Crude Framework for Bypassing Exposure," *Journal of Safety Research*, Vol. 2, pp. 26-29.

Haight, F.A. (1973). "Induced Exposure," *Accident Analysis and Prevention*, Vol. 5, pp. 111-126.

IIHS Advisory No. 5, July 1988, http://www.iihs.org/research/advisories/iihs_advisory_5.pdf.

Kahane, C.J. (1994). *Correlation of NCAP Performance with Fatality Risk in Actual Head-On Collisions*, NHTSA Technical Report No. DOT HS 808 061. Washington, DC: National Highway Traffic Safety Administration, http://www-nrd.nhtsa.dot.gov/Pubs/808061.PDF.

Kahane, C. J. (1997). *Relationships Between Vehicle Size and Fatality Risk in Model Year 1985-93 Passenger Cars and Light Trucks*, NHTSA Technical Report. DOT HS 808 570. Washington, DC: National Highway Traffic Safety Administration, http://www-nrd.nhtsa.dot.gov/Pubs/808570.PDF .

Kahane, C. J. (2003). *Vehicle Weight, Fatality Risk and Crash Compatibility of Model Year 1991-99 Passenger Cars and Light Trucks*, NHTSA Technical Report. DOT HS 809 662. Washington, DC: National Highway Traffic Safety Administration, http://www-nrd.nhtsa.dot.gov/Pubs/809662.PDF.

Kahane, C. J. (2004). *Lives Saved by the Federal Motor Vehicle Safety Standards and Other Vehicle Safety Technologies, 1960-2002*, NHTSA Technical Report No. DOT HS 809 833. Washington, DC: National Highway Traffic Safety Administration, http://www-nrd.nhtsa.dot.gov/Pubs/809833.PDF.

Kahane, C. J. (2007). *An Evaluation of Side Impact Protection – FMVSS 214 TTI(d) Improvements and Side Air Bags*, NHTSA Technical Report No. DOT HS 810 748. Washington, DC: National Highway Traffic Safety Administration, http://www-nrd.nhtsa.dot.gov/Pubs/810748.PDF.

Kahane, C. J. (2010). "Relationships Between Fatality Risk, Mass, and Footprint in Model Year 1991-1999 and Other Passenger Cars and LTVs," *Final Regulatory Impact Analysis: Corporate Average Fuel Economy for MY 2012-MY 2016 Passenger Cars and Light Trucks*. Washington, DC: National Highway Traffic Safety Administration, pp. 464-542, http://www.nhtsa.dot.gov/staticfiles/DOT/NHTSA/Rulemaking/Rules/Associated%20Files/CAFE_2012-2016_FRIA_04012010.pdf

Kahane, C. J. (2011). *Relationships Between Fatality Risk, Mass, and Footprint in Model Year 2000-2007 Passenger Cars and LTVs – Preliminary Report*. (Docket No. NHTSA-2010-0152-0023). Washington, DC: National Highway Traffic Safety Administration.

Kullgren, A., Lie, A., and Tingvall, C. (2010). "Comparison Between Euro NCAP Test Results and Real-World Crash Data," *Traffic Injury Prevention*, Vol. 11, pp. 587-593.

Light-Duty Vehicle Greenhouse Gas Emission Standards and Corporate Average Fuel Economy Standards; Final Rule. 75 Fed. Reg. 25324 (May 7, 2010)

MacLaughlin, T.F., and Kessler, J.W. (1990). *Pedestrian Head Impact Against the Central Hood of Motor Vehicles – Test Procedure and Results*. Paper No. 902315. Warrendale, PA: Society of Automotive Engineers.

Mela, D. F. (1974). "How Safe Can We Be in Small Cars?" *International Congress on Automotive Safety, 3rd,* NHTSA Technical Report. DOT HS 801 481. Washington, DC: National Highway Traffic Safety Administration.

Menard, B., ed. (2012). *Peer Review of LBNL Statistical Analysis of the Effect of Vehicle Mass & Footprint Reduction on Safety (LBNL Phase 1 and 2 Reports)* (To appear in Docket No. NHTSA-2010-0152). Charlottesville, VA: Systems Research and Application Corp.

Najm, W.G., Sen, B., Smith, J.D., and Campbell, B.N. (2003). *Analysis of Light Vehicle Crashes and Pre-Crash Scenarios Based on the 2000 General Estimates System,* Report No. DOT HS 809 573. Washington, DC: National Highway Traffic Safety Administration.

NAS (1996). Peer-review letter from D. Warner North to Ricardo Martinez, NHTSA, July 12, 1996. Washington, DC: National Research Council.

NAS (2002). *Effectiveness and Impact of Corporate Average Fuel Economy (CAFE) Standards.* Washington, DC: National Research Council.

New Crash Tests Demonstrate the Influence of Vehicle Size and Weight on Safety in Crashes, IIHS News Release, April 14, 2009, http://www.iihs.org/news/rss/pr041409.html.

News Release, February 24, 1998, http://www.iihs.org/news/1998/iihs_news_022498.pdf.

NHTSA (1991). *Effect of Car Size on Fatality and Injury Risk*. Washington, DC: National Highway Traffic Safety Administration.

NHTSA (2000). *Traffic Safety Facts 1999.* Report No. DOT HS 809 100. Washington, DC: National Highway Traffic Safety Administration.

NHTSA (2003). *Memorandum: Drs. James H. Hedlund, Adrian K. Lund and Donald W. Reinfurt's Reviews and Comments of the Draft Technical Report*. (Docket No. NHTSA-2003-16318-0004). Washington, DC: National Highway Traffic Safety Administration.

NHTSA (2010). *Traffic Safety Facts 2009.* Report No. DOT HS 811 402. Washington, DC: National Highway Traffic Safety Administration, http://www-nrd.nhtsa.dot.gov/Pubs/811402.pdf.

Partyka, S.C. (1995). *Impacts with Yielding Fixed Objects by Vehicle Weight*. NHTSA Technical Report. DOT HS 808 574. Washington, DC: National Highway Traffic Safety Administration.

Robertson, L.S. (1991), "How to Save Fuel and Reduce Injuries in Automobiles," The Journal of Trauma, Vol. 31, pp. 107-109.

Schadler, A. Multicollinearity in Logistic Regression. Lexington, KY: University of Kentucky Center for Statistical Computing Support. http://www.uky.edu/ComputingCenter/SSTARS/MulticollinearityinLogisticRegression.htm.

Shen, J., and Gao, S. (2008). "A Solution to Separation and Multicollinearity in Multiple Logistic Regression," *Journal of Data Science*, Vol. 6, pp. 515-531.

Sivinski R. (2011). *Update of NHTSA's 2007 Evaluation of the Effectiveness of Light Vehicle Electronic Stability Control (ESC) in Crash Prevention*, NHTSA Technical Report No. DOT HS 811 486. Washington, DC: National Highway Traffic Safety Administration. http://www-nrd.nhtsa.dot.gov/Pubs/811486.pdf.

Stutts, J. C., and Martell, C. (1992), "Older Driver Population and Crash Involvement Trends, 1974-1988," *Accident Analysis and Prevention*, Vol. 28, pp. 317-327.

Table MV-1, State Motor-Vehicle Registrations – 2008, http://www.fhwa.dot.gov/policyinformation/statistics/2008/xls/mv1.xls.

Thorpe, J.D. (1964), "Calculating Relative Involvement Rates in Accidents without Determining Exposure," *Australian Road Research*, Vol. 2, pp. 25-36.

Vago, E., and Kemeny, S. (2006). "Logistic Ridge Regression for Clinical Data Analysis (A Case Study)," *Applied Ecology & Environmental Research*, Vol. 4, pp.171-179.

Van Auken, R. M., and Zellner, J. W. (2003). A Further Assessment of the Effects of Vehicle Weight and Size Parameters on Fatality Risk in Model Year 1985-98 Passenger Cars and 1986-97 Light Trucks. Report No. DRI-TR-03-01. Torrance, CA: Dynamic Research, Inc.

Van Auken, R. M., and Zellner, J. W. (2005a). An Assessment of the Effects of Vehicle Weight and Size on Fatality Risk in 1985 to 1998 Model Year Passenger Cars and 1985 to 1997 Model Year Light Trucks and Vans. Paper No. 2005-01-1354. Warrendale, PA: Society of Automotive Engineers.

Van Auken, R. M., and Zellner, J. W. (2005b). Supplemental Results on the Independent Effects of Curb Weight, Wheelbase, and Track on Fatality Risk in 1985-1998 Model Year Passenger Cars and 1986-97 Model Year LTVs. Report No. DRI-TR-05-01. Torrance, CA: Dynamic Research, Inc.

Van Auken, R.M., and Zellner, J. W. (2012a). *Updated Analysis of the Effects of Passenger Vehicle Size and Weight on Safety, Phase I*. Report No. DRI-TR-11-01. (Docket No. NHTSA-2010-0152-0030). Torrance, CA: Dynamic Research, Inc.

Van Auken, R.M., and Zellner, J. W. (2012b). *Updated Analysis of the Effects of Passenger Vehicle Size and Weight on Safety, Phase II; Preliminary Analysis Based on 2002 to 2008 Calendar Year Data for 2000 to 2007 Model Year Light Passenger Vehicles to Induced-Exposure and Vehicle Size Variables*. Report No. DRI-TR-12-01, Vols. 1-3. (Docket No. NHTSA-2010-0152-0032). Torrance, CA: Dynamic Research, Inc.

Van Auken, R.M., and Zellner, J. W. (2012c). *Updated Analysis of the Effects of Passenger Vehicle Size and Weight on Safety, Phase II; Preliminary Analysis Based on 2002 to 2008 Calendar Year Data for 2000 to 2007 Model Year Light Passenger Vehicles to Induced-Exposure and Vehicle Size Variables*. Report No. DRI-TR-12-01, Vols. 4-5. (Docket No. NHTSA-2010-0152-0033). Torrance, CA: Dynamic Research, Inc.

Van Auken, R.M., and Zellner, J. W. (2012d). *Updated Analysis of the Effects of Passenger Vehicle Size and Weight on Safety; Sensitivity of the Estimates for 2002 to 2008 Calendar Year Data for 2000 to 2007 Model Year Light Passenger Vehicles to Induced-Exposure and Vehicle Size Variables.* Report No. DRI-TR-12-03. (Docket No. NHTSA-2010-0152-0034). Torrance, CA: Dynamic Research, Inc.

Van Der Zwaag, D.D. (1971), "Induced Exposure as a Tool to Determine Passenger Car and Truck Involvement in Accidents," *HIT Lab Reports*, Vol. 1, pp. 1-8.

Ward's Automotive Yearbook. Annual publication. Southfield, MI: Penton Media, Inc.

Wenzel, T. (2011). *Assessment of NHTSA's Report "Relationships Between Fatality Risk, Mass, and Footprint in Model Year 2000-2007 Passenger Cars and LTVs – Draft Final Report."* (Docket No. NHTSA-2010-0152-0026). Berkeley, CA: Lawrence Berkeley National Laboratory.

Wenzel, T. (2012). *Assessment of NHTSA's Report "Relationships Between Fatality Risk, Mass, and Footprint in Model Year 2000-2007 Passenger Cars and LTVs – Final Report."* (To appear in Docket No. NHTSA-2010-0152). Berkeley, CA: Lawrence Berkeley National Laboratory.

Appendix A

MY 2000-2007 SUVs Considered Crossover Utility Vehicles (CUVs) in this Report

Ward's Automotive Yearbooks® began to list certain makes and models as "CUVs" in MY 2002 in its "Light-Duty Truck Specifications" and its narrative descriptions of the new vehicles.[164] In general, the list of vehicles considered CUVs in this report includes all those called CUVs by Ward's but also some other vehicles (among them, any CUVs of designs discontinued before 2002, since Ward's was not yet using the term), based on sources or criteria documented in the footnotes.[165]

	First MY[166]	Last MY[167]	LTV Groups Included (NHTSA 5-Digit Codes)
Chrysler PT Cruiser	2001	2007	6304
Chrysler Pacifica	2004	2007	6307
Dodge Magnum	2005	2007	6310
Jeep Compass/Patriot	2007	2007	6312
Ford Escape/Merc Mariner/Mazda Tribute	2001	2007	12309
Ford Freestyle	2005	2007	12312
Ford Edge/Lincoln MKX	2007	2007	12314
Pontiac Aztec	2001	2005	18313
Buick Rendezvous	2002	2007	18315
Saturn Vue	2002	2007	18317
Cadillac SRX	2004	2007	18319
Chevrolet HHR	2006	2007	18321
GMC Acadia/Saturn Outlook	2007	2007	18323
VW Touareg/Porsche Cayenne	2004	2007	30301
Audi Allroad	2001	2005	32301
Audi Q7	2007	2007	32302
BMW X5	2000	2007	34301, 34303
BMW X3	2004	2007	34302
Nissan Murano	2003	2007	35304
Infiniti FX35/45	2003	2007	35305
Honda CR-V	1997	2007	37301, 37303, 37307
Acura MDX	2001	2007	37302, 37305
Honda Pilot	2003	2007	37302

---

[164] Southfield, MI: Penton Media, Inc.
[165] Ward's does not call the Jeep Liberty a CUV, nor the Dodge Nitro (a related design with a longer wheelbase, according to cars.com). Although these vehicles are of unibody design, they have additional structure for the front and rear suspension to provide better off-road and towing capabilities and for that reason it is probably more appropriate to consider them truck-based SUVs.
[166] But only the vehicles of MY 2000 and onward are included in the analyses of this report.
[167] 2007 is the last model year in this report; however, most of these vehicles continued as CUVs beyond MY 2007.

	First MY[168]	Last MY[169]	LTV Groups Included (NHTSA 5-Digit Codes)
Honda Element	2003	2007	37304
Acura RDX	2007	2007	37306
Mazda CX-7	2007	2007	41301
Mazda CX-9	2007	2007	41302
Mercedes ML[170]	2006	2007	42303
Mercedes R[171]	2006	2007	42304
Mercedes GL	2007	2007	42305
Subaru Forester	1998	2007	48301
Subaru Outback[172]	2005	2007	48302
Subaru Tribeca	2006	2007	48303
Subaru Baja	2003	2006	48702
Toyota RAV4[173]	1996	2007	49306, 49309, 49313
Lexus RX300	1999	2003	49308
Toyota Highlander	2001	2007	49310
Lexus RX330/350	2004	2007	49310
Volvo XC70	2001	2007	51301
Volvo XC90	2003	2007	51302
Mitsubishi Outlander[174]	2003	2007	52305, 52307
Mitsubishi Endeavor[175]	2004	2007	52306
Chevrolet Equinox/Pontiac Torrent	2005	2007	53307
Suzuki XL-7[176]	2007	2007	53309
Hyundai Santa Fe	2001	2007	55301, 55304
Hyundai Tucson/Kia Sportage[177]	2005	2007	55302
Hyundai Veracruz	2007	2007	55303
Land Rover Freelander	2002	2005	62307

---

[168] But only the vehicles of MY 2000 and onward are included in the analyses of this report.

[169] 2007 is the last model year in this report; however, most of these vehicles continued as CUVs beyond MY 2007.

[170] But not the 1998-2005 Mercedes ML (LTV group 42301). Mercedes ML was redesigned in 2006 as a unibody and it is a crossover according to http://en.wikipedia.org/wiki/Crossover_(automobile), (even though Ward's does not call it a CUV).

[171] Mercedes R was built on the same platform as ML starting in 2006 and it is a crossover according to msn.auto, http://editorial.autos.msn.com/article.aspx?cp-documentid=435375, (even though Ward's does not call it a CUV).

[172] But not the 2003-2004 Subaru Outback (car group 48013), which is a passenger car by NHTSA's definitions for FMVSS and CAFE purposes. In 2005, Subaru Outback was reclassified an LTV and became a CUV (even though Ward's still calls it a car).

[173] http://en.wikipedia.org/wiki/Toyota_RAV4 says RAV4 was the first crossover SUV, based on Corolla.

[174] Ward's does not call it a CUV until 2007, but http://en.wikipedia.org/wiki/Mitsubishi_Outlander calls it a CUV from the start.

[175] Ward's does not call it a CUV until 2007, but cars.com calls it a CUV from the start.

[176] Ward's states that the 2007 XL-7 (LTV group 53309) was redesigned as a CUV, whereas the 2001-2006 XL-7 (LTV group 53306) was truck-based.

[177] Ward's does not call them CUVs, but http://en.wikipedia.org/wiki/Kia_Sportage calls them CUVs based on the Elantra platform (whereas the 1995-2002 Sportage, LTV groups 63301/63302, was based on the Mazda Bongo van and was not a CUV).

Appendix B

Codebook for the Database of Fatal Crash Involvements

Observations: 113,248
Variables: 85

Alphabetic List of Variables

ABS – Antilock brake system (4-wheel)   Type: Numeric8
     0   Not ABS-equipped
     .01-.99   proportion of vehicles with ABS (optional but cannot be identified for an individual vehicle from the first 12 digits of its VIN)
     1   ABS-equipped

AWD – All-wheel-drive or 4-wheel-drive   Type: Numeric8
     0   FWD or RWD – i.e., not equipped with AWD or 4-wheel-drive
     .01-.99   proportion of vehicles with AWD or 4-wheel-drive (optional but cannot be identified from first 12 digits of VIN)
     1   equipped with AWD or 4-wheel-drive

BLOCKER – Voluntary vehicle-to-vehicle compatibility certification   Type: Numeric8
     0   Not certified (or not a pickup truck or SUV)
     1   Certified, Option 1 (bumper height overlap with passenger cars)
     2   Certified, Option 2 ("blocker beam" or other secondary energy-absorbing structure)

BOD2 – Passenger car body type   Type: Numeric8
     0   Not a passenger car
     1   2-door convertible
     2   2-door coupe or sedan
     3   3-door hatchback
     4   4-door sedan
     5   5-door hatchback
     6   Station wagon

CARS – Number of passenger cars in the crash   Type: Numeric8
     0 – 44   Number of passenger cars in transport involved in the crash

CG – Car group   Type: Numeric8
     1303-64005   Download 10Formats2007.sas, CarGroup2007.docx, or LTVGroup2007.docx for valid codes

CGP – Car group codes for merging with Polk data   Type: Numeric8
   1303-64005   Download 10Formats2007.sas, CarGroup2007.docx, or LTVGroup2007.docx for valid codes

COMBO – Side air bag with torso/head protection ("combination" bag)   Type: Numeric8
   0   Not equipped with combo bags
   .01-.99   proportion of vehicles with combo bags (optional but cannot be identified from first 12 digits of VIN)
   .75   Combo bag for driver only
   1   Combo bags for driver and RF passenger (Note: a combo bag is a single bag that protects the head and torso; separate curtains and torso bags are coded CURTAIN=1, TORSO=1, COMBO=0)

COUNTY – County FIPS code   Type: Numeric3
   1 – 840   3-digit FIPS code for the county (unmodified FARS variable)

CRSH – Crash type   Type: Numeric8
   1   first-event rollover
   2   Hit fixed object
   3   Hit pedestrian/bike/motorcycle
   4   Hit heavy vehicle (i.e., GVWR ≥ 10,000 pounds)
   5   Hit passenger car of known mass
   6   Hit LTV of known mass and GVWR < 10,000 pounds
   11   Other non-collision (fire, immersion, fell from vehicle,…)
   12   Hit train
   13   Hit animal, working vehicle, or on-road object
   14   Single-vehicle crash, no non-occupants: other/unclear type
   15   Single-vehicle crash involving non-occupants: other/unclear type
   16   Single vehicle: fatal to multiple or other traffic units
   21   Hit car of unknown mass
   22   Hit LTV of unknown mass
   23   Hit "other" type of vehicle (snowmobile, farm equipment, construction machinery,…)
   24   Hit unknown type of vehicle
   25   2-vehicle crash: fatal to non-occupants or parked-vehicle-occupants in addition to, possibly, occupants of vehicles in transport
   31   3+ vehicle crash: fatal only to occupants of vehicles in transport
   32   3+ vehicle crash: fatal to non-occupants or parked-vehicle-occupants in addition to, possibly, occupants of vehicles in transport

CURBWT – Curb weight (pounds)   Type: Numeric8
   1799 – 7520   Curb weight in pounds

CURTAIN – Head curtain air bags for front-seat occupants   Type: Numeric8
    0   Not equipped with head curtain bags
    .01-.99   proportion of vehicles with curtain bags (optional but cannot be identified from first 12 digits of VIN)
    1   Head curtains for driver and RF passenger

CY – Calendar year   Type: Numeric8
    2002 – 2008   Calendar year in which the crash occurred

DEATHS – Case vehicle occupant fatalities   Type: Numeric3
    0 – 14   Number of occupant fatalities in the case vehicle (unmodified FARS variable)

DENS3 – County population density   Type: Numeric8
    0.1-66951   Inhabitants per square mile in the county where the crash occurred, based on 2000 census

DRVAGE – Driver age   Type: Numeric8
    14 – 96   Age of the driver (unknown or out-of-range excluded)

DRVMALE – Male driver   Type: Numeric8
    0   Female driver
    1   Male driver (non-reported gender excluded)

ESC – Electronic stability control   Type: Numeric8
    0   Not ESC-equipped
    .01-.99   proportion of vehicles with ESC (optional but cannot be identified for an individual vehicle from the first 12 digits of its VIN)
    1   ESC-equipped

FATALS – Fatalities in the crash   Type: Numeric3
    0 – 14   Total number of fatalities in the crash, including occupants of any vehicles (in transport or not-in-transport) and non-occupants (unmodified FARS variable)

FOOTPRNT – Footprint (square feet)   Type: Numeric8
    34.6 – 81.0   Footprint in square feet (TRAKWDTH X WB_MIN converted to square feet)

GE10 – GVWR $\geq$ 10,000 pounds   Type: Numeric8
    0   GVWR (gross vehicle weight rating) < 10,000 pounds
    1   GVWR $\geq$ 10,000 pounds

HARM_EV – First harmful event   Type: Numeric3
    1 – 99   First harmful event in the crash (unmodified FARS variable)

HIFAT_ST – High-fatality-rate State   Type: Numeric8
   0   One of the 26 States with < 160 crash fatalities per million vehicle registration years in 2002-2008
   1   One of the 24 States with > 160 crash fatalities per million vehicle registration years in 2002-2008

HOUR – Hour when the crash occurred   Type: Numeric3
   0 – 24   Hour when the crash occurred, military time (unmodified FARS variable)

HVYTRKS – Number of heavy vehicles in the crash   Type: Numeric8
   0 – 28   Number of heavy vehicles in transport involved in the crash

LGT_COND – Light condition   Type: Numeric3
   1 – 9   Light condition at the time of the crash (unmodified FARS variable)

LTVS – Number of LTVs in the crash   Type: Numeric8
   0 – 51   Number of LTVs in transport involved in the crash

MAK2 – Vehicle Make   Type: Numeric8
   2   Jeep
   3   Hummer
   6   Chrysler
   7   Dodge
   9   Plymouth
   11   Sprinter
   12   Ford
   13   Lincoln
   14   Mercury
   18   Buick
   19   Cadillac
   20   Chevrolet
   21   Oldsmobile
   22   Pontiac
   23   GMC
   24   Saturn
   30   Volkswagen
   32   Audi
   33   Mini-Cooper
   34   BMW
   35   Nissan
   37   Honda
   38   Isuzu
   39   Jaguar
   41   Mazda
   42   Mercedes-Benz
   45   Porsche

47	Saab
48	Subaru
49	Toyota (including Scion)
51	Volvo
52	Mitsubishi
53	Suzuki
54	Acura
55	Hyundai
58	Infiniti
59	Lexus
62	Land-Rover
63	Kia
64	Daewoo

MAXVEHNO – Highest VEH_NO of vehicles hitting non-occupants   Type: Numeric8
    1 – 99   Highest VEH_NO of the vehicles that struck and fatally injured a non-occupant

MCYCLES – number of motorcycles in the crash   Type: Numeric8
    0 – 6   Number of motorcycles in transport involved in the crash

MINVEHNO – Lowest VEH_NO of vehicles hitting non-occupants   Type: Numeric8
    1 – 99   Lowest VEH_NO of the vehicles that struck and fatally injured a non-occupant

MM2 – Make-model   Type: Numeric8
    2001-64033   Download 10Formats2007.sas, CarGroup2007.docx, or LTVGroup2007.docx for valid codes

MMP – Make-model codes for merging with Polk data   Type: Numeric8
    2001-64033   Download 10Formats2007.sas, CarGroup2007.docx, or LTVGroup2007.docx for valid codes

MY – Model year   Type: Numeric8
    2000 – 2007   Model year of the case vehicle

M_HARM – Most harmful event   Type: Numeric3
    1 – 99   Most harmful event for the case vehicle (unmodified FARS variable)

NITE – Time of day when the crash occurred   Type: Numeric8
    0   6:00 a.m. to 6:59 p.m.
    1   7:00 p.m. to 5:59 a.m.

NONOCC – Number of non-occupant fatalities in the crash   Type: Numeric8
    0 – 10   Number of non-occupant fatalities in the crash

OBODY – Body type of the other vehicle   Type: Numeric8
    1 – 99   Body type of the other vehicle in a 2-vehicle crash (unmodified FARS variable BODY_TYP for the other vehicle)

OCC – Number of occupant fatalities in the crash   Type: Numeric8
    0 – 14   Number of occupant fatalities in the crash (including occupants of other vehicles in transport)

OCG2 – Car group (or LTV group) of the other vehicle   Type: Numeric8
    1008-99999   5-digit VIN-derived vehicle group (CG) of the other vehicle in a 2-vehicle crash (Download 10Formats2007.sas, CarGroup2007.docx, or LTVGroup2007.docx for valid codes)

OCURBWT – Curb weight (pounds) of the other vehicle   Type: Numeric8
    190 – 7520   Curb weight in pounds of the other vehicle in a 2-vehicle crash

OMAKMOD – FARS make and model of the other vehicle   Type: Numeric8
    1005-99999   Make and model of the other vehicle in a 2-vehicle crash (unmodified FARS variable MAK_MOD for the other vehicle)

OMM2 – Make and model of the other vehicle   Type: Numeric8
    1005-99999   5-digit VIN-derived make-model (MM2) of the other vehicle in a 2-vehicle crash (Download 10Formats2007.sas, CarGroup2007.docx, or LTVGroup2007.docx for valid codes)

OMOD_YR – Model year of the other vehicle   Type: Numeric8
    1930 – 9999   Model year of the other vehicle in a 2-vehicle crash (unmodified FARS variable MOD_YEAR for the other vehicle)

OTHVEH – Number of other-type vehicles in the crash   Type: Numeric8
    0 – 1   Number of other-type vehicles (snowmobiles, farm equipment, …) in transport involved in the crash

OVIN – VIN of the other vehicle   Type: Character12
    12-character VIN of the other vehicle in a 2-vehicle crash (unmodified FARS variable VIN for the other vehicle)

OVINA – VINA_MOD of the other vehicle   Type: Character3
    VINA_MOD of the other vehicle in a 2-vehicle crash (unmodified FARS variable VINA_MOD for the other vehicle)

OVTYP – Type of the other vehicle in a 2-vehicle crash   Type: Numeric8
- 1   Passenger car
- 2   LTV
- 3   Heavy vehicle
- 4   Motorcycle
- 5   Other (snowmobile, farm equipment, construction machinery,…)
- 9   Unknown

OVTYP2 – Type (detailed) of the other vehicle in a 2-vehicle crash   Type: Numeric8
- 1   Passenger car
- 2.1   CUV
- 2.2   Minivan
- 2.3   Astro/Safari/Aerostar
- 2.4   Truck-based LTV
- 3   Heavy vehicle
- 4   Motorcycle
- 5   Other (snowmobile, farm equipment, construction machinery,…)
- 9   Unknown

OWTFLAG – Source of curb weight for the other vehicle   Type: Numeric8
- 0   Good VIN, MY 1985-2007, weight in database
- 1   Good VIN, filled in some gaps in MY 1985-1999 database
- 2   MM2 not defined, but FARS supplies a VIN_WGT
- 3   MY 1981-1984 car with MM2 decoded, use FARS VIN_WGT
- 4   Weights in database for RV cutaways, etc. excluded body; reset to 5000
- 5   Good VIN, MY 1981-1984 or 2008-2009, use 1985 or 2007 weights
- 6   Filled in approximate weight based on MAK_MOD
- 7   Car or LTV, could not define a weight
- 9   Not a car or LTV

PARK – Number of non-transport-vehicle-occupant fatalities   Type: Numeric8
- 0 – 4   Number of non- transport-vehicle-occupant fatalities in the crash

PASSIVE – Frontal air bags   Type: Numeric8
- 0   Not equipped with frontal air bags
- 2   Dual frontal air bags
- 3   Dual frontal air bags with a manual on-off switch for the RF passenger
- 9   Unknown if equipped and/or what type

RDSUR – Road surface condition   Type: Numeric8
- 0   Dry
- 1   Wet, muddy, or oily
- 2   Snow, ice, or slush
- 9   Unknown

ROAD_FNC – Roadway function class   Type: Numeric3
    1 –99   Roadway function class (unmodified FARS variable)

ROLLCURT – Head curtain air bags designed to deploy in rollovers   Type: Numeric8
    0   Not equipped with rollover curtain bags
     .01-.99   proportion of vehicles with rollover curtain bags (optional but cannot be identified from first 12 digits of VIN)
    1   Rollover curtains for driver and RF passenger (Note: CURTAIN should also be coded 1)

RURAL – County population density   Type: Numeric8
    0   Crash occurred in a county with ≥ 250 inhabitants per square mile in the 2000 census
    1   Crash occurred in a county with < 250 inhabitants per square mile

SPDLIM55 – Speed limit 55+   Type: Numeric8
    0   Speed limit < 55 mph or not reported, for all roadways involved in the crash
    1   Speed limit ≥ 55 mph for at least one roadway involved in the crash

SP_LIMIT – Speed limit   Type: Numeric3
    0 – 99   Highest speed limit of roadways involved in the crash (unmodified FARS variable)

SQUADCAR – Police model   Type: Numeric8
    0   Not a Ford Crown Victoria or Chevrolet Impala "police" model and not a 2004-2005 Chevrolet Impala SS
    1   Ford Crown Victoria or Chevrolet Impala "police" model or 2004-2005 Chevrolet Impala SS

STATE – State FIPS code   Type: Numeric3
    1 – 56   2-digit FIPS code for the State (unmodified FARS variable)

ST_CASE – State-case ID number   Type: Numeric4
10001–560168   State-case ID number (unmodified FARS variable)

SUR_COND – Roadway surface condition   Type: Numeric3
    1 –9   Roadway surface condtition (unmodified FARS variable)

TORSO – Side air bag with torso protection   Type: Numeric8
    0   Not equipped with torso bags
     .01-.99   proportion of vehicles with torso bags (optional but cannot be identified from first 12 digits of VIN)
     .25   Torso bag for RF passenger only
    1   Torso bags for driver and RF passenger

TRAKWDTH – Track width (inches, average of front and rear wheels)   Type: Numeric8
    54 – 72.65   Average of front and rear track width, in inches

TRKTYP – LTV type (case vehicle)   Type: Numeric8
    0   Passenger car
    1   Compact pickup truck (download LTVGroup2007.docx for more information)
    2   Full-sized pickup truck with GVWR < 10,000 pounds
    3   Compact SUV
    4   Full-sized SUV
    5   Minivan
    6   Full-sized van with GVWR < 10,000 pounds
    7   Pickup-car (e.g., Subaru Baja)
    12   300-series pickup truck with GVWR ≥ 10,000 pounds
    16   300-series van with GVWR ≥ 10,000 pounds

TW_F – Track width, front wheels (inches)   Type: Numeric8
    54.5 – 71.6   Front track width, in inches

TW_R – Track width, rear wheels (inches)   Type: Numeric8
    52.2 – 75.8   Rear track width, in inches

UNKVEH – Number of unknown-type vehicles in the crash   Type: Numeric8
    0 – 5   Number of unknown-type vehicles in transport involved in the crash

V1 – 1st character of the VIN   Type: Character1
V2 – 2nd character of the VIN   Type: Character1
V3 – 3rd character of the VIN   Type: Character1
V4 – 4th character of the VIN   Type: Character1
V5 – 5th character of the VIN   Type: Character1
V6 – 6th character of the VIN   Type: Character1
V7 – 7th character of the VIN   Type: Character1
V8 – 8th character of the VIN   Type: Character1
V11 – 11th character of the VIN   Type: Character1
V12 – 12th character of the VIN   Type: Character1

VEHAGE – Vehicle age in years (CY – MY)   Type: Numeric8
    0 – 8   Age of the case vehicle (years)

VEH_NO – Vehicle ID number   Type: Numeric3
    1 – 91   Vehicle ID number (unmodified FARS variable)

VE_FORMS – Number of vehicle forms submitted   Type: Numeric3
    1 – 92   Number of vehicle-in-transport forms submitted for this crash (unmodified FARS variable)

VINA_MOD   Type: Character3
    3-character make-model code (unmodified FARS variable)

VTYP – Case vehicle type   Type: Numeric8
    1   Passenger car, 2 doors
    2   Passenger car, 4 doors
    3   Pickup truck, light duty (compact or 150-series)
    4   Pickup truck, heavy duty (250- or 350-series)
    5   SUV, truck-based
    6   CUV (crossover SUV)
    7   Minivan, except Chevrolet Astro or GMC Safari
    7.1   Chevrolet Astro or GMC Safari
    8   Full-sized van

WB_MAX – Maximum wheelbase (inches, for selected pickup trucks)   Type: Numeric8
    117.5 – 172.4 Maximum wheelbase for pickup trucks that can have 2 or more bed lengths for the same 12-character VIN (left blank for all other vehicles and not used in computing footprint)

WB_MIN – Wheelbase (inches)   Type: Numeric8
    86.6 – 170.3 Wheelbase (and if this is a pickup truck, wheelbase for the shortest bed available for this 12-character VIN)

Appendix C

Codebook for the Database of Induced-Exposure Crash Involvements
(Crashes from 13 State files, with national weights)

Observations: 2,457,228
Variables: 49

Alphabetic List of Variables

ABS – Antilock brake system (4-wheel) Type: Numeric8
- 0   Not ABS-equipped
- .01-.99   proportion of vehicles with ABS (optional but cannot be identified for an individual vehicle from the first 12 digits of its VIN)
- 1   ABS-equipped

AWD – All-wheel-drive or 4-wheel-drive Type: Numeric8
- 0   FWD or RWD – i.e., not equipped with AWD or 4-wheel-drive
- .01-.99   proportion of vehicles with AWD or 4-wheel-drive (optional but cannot be identified from first 12 digits of VIN)
- 1   equipped with AWD or 4-wheel-drive

BLOCKER – Voluntary vehicle-to-vehicle compatibility certification Type: Numeric8
- 0   Not certified (or not a pickup truck or SUV)
- 1   Certified, Option 1 (bumper height overlap with passenger cars)
- 2   Certified, Option 2 ("blocker beam" or other secondary energy-absorbing structure)

BOD2 – Passenger car body type Type: Numeric8
- 0   Not a passenger car
- 1   2-door convertible
- 2   2-door coupe or sedan
- 3   3-door hatchback
- 4   4-door sedan
- 5   5-door hatchback
- 6   Station wagon

CG – Car group Type: Numeric8
- 1303-64005   Download 10Formats2007.sas, CarGroup2007.docx, or LTVGroup2007.docx for valid codes

CGP – Car group codes for merging with Polk data Type: Numeric8
- 1303-64005   Download 10Formats2007.sas, CarGroup2007.docx, or LTVGroup2007.docx for valid codes

COMBO – Side air bag with torso/head protection ("combination" bag)   Type: Numeric8
- 0   Not equipped with combo bags
- .01-.99   proportion of vehicles with combo bags (optional but cannot be identified from first 12 digits of VIN)
- .75   Combo bag for driver only
- 1   Combo bags for driver and RF passenger (Note: a combo bag is a single bag that protects the head and torso; separate curtains and torso bags are coded CURTAIN=1, TORSO=1, COMBO=0)

CURBWT – Curb weight (pounds)   Type: Numeric8
- 1799 – 7520   Curb weight in pounds

CURTAIN – Head curtain air bags for front-seat occupants   Type: Numeric8
- 0   Not equipped with head curtain bags
- .01-.99   proportion of vehicles with curtain bags (optional but cannot be identified from first 12 digits of VIN)
- 1   Head curtains for driver and RF passenger

CY – Calendar year   Type: Numeric8
- 2002 – 2008   Calendar year in which the crash occurred

DRVAGE – Driver age   Type: Numeric8
- 14 – 96   Age of the driver (unknown or out-of-range excluded)

DRVMALE – Male driver   Type: Numeric8
- 0   Female driver
- 1   Male driver (non-reported gender excluded)

DUMMY – "Dummy" induced-exposure case   Type: Numeric8
- 0   Original induced-exposure case, same MY and CY as the case vehicle
- 1   No original case available, used a case of the same make-model and MY, but different CY
- 2   No original case available, used a case of the same make-model, but different MY and CY

ESC – Electronic stability control   Type: Numeric8
- 0   Not ESC-equipped
- .01-.99   proportion of vehicles with ESC (optional but cannot be identified for an individual vehicle from the first 12 digits of its VIN)
- 1   ESC-equipped

FOOTPRNT – Footprint (square feet)   Type: Numeric8
- 34.6 – 81.0   Footprint in square feet (TRAKWDTH X WB_MIN converted to square feet)

GE10 – GVWR ≥ 10,000 pounds   Type: Numeric8
     0   GVWR (gross vehicle weight rating) < 10,000 pounds
     1   GVWR ≥ 10,000 pounds

HIFAT_ST – High-fatality-rate State   Type: Numeric8
     0   One of the 26 States with < 160 crash fatalities per million vehicle registration years in 2002-2008
     1   One of the 24 States with > 160 crash fatalities per million vehicle registration years in 2002-2008

MAK2 – Vehicle Make   Type: Numeric8
     2   Jeep
     3   Hummer
     6   Chrysler
     7   Dodge
     9   Plymouth
     11   Sprinter
     12   Ford
     13   Lincoln
     14   Mercury
     18   Buick
     19   Cadillac
     20   Chevrolet
     21   Oldsmobile
     22   Pontiac
     23   GMC
     24   Saturn
     30   Volkswagen
     32   Audi
     33   Mini-Cooper
     34   BMW
     35   Nissan
     37   Honda
     38   Isuzu
     39   Jaguar
     41   Mazda
     42   Mercedes-Benz
     45   Porsche
     47   Saab
     48   Subaru
     49   Toyota (including Scion)
     51   Volvo
     52   Mitsubishi
     53   Suzuki
     54   Acura
     55   Hyundai

58	Infiniti
59	Lexus
62	Land-Rover
63	Kia
64	Daewoo

MM2 – Make-model   Type: Numeric8
 2001-64033 Download 10Formats2007.sas, CarGroup2007.docx, or LTVGroup2007.docx for valid codes

MMP – Make-model codes for merging with Polk data Type: Numeric8
 2001-64033 Download 10Formats2007.sas, CarGroup2007.docx, or LTVGroup2007.docx for valid codes

MY – Model year   Type: Numeric8
 2000 – 2007 Model year of the case vehicle

NITE – Time of day when the crash occurred Type: Numeric8
 0   6:00 a.m. to 6:59 p.m.
 .01-.99 In 2002-2006 Alabama data, estimated probability that the crash occurred from 7:00 p.m. to 5:59 a.m., depending on light condition and month (as inferred from case number)
 1   7:00 p.m. to 5:59 a.m.

PASSIVE – Frontal air bags   Type: Numeric8
 0 Not equipped with frontal air bags
 2 Dual frontal air bags
 3 Dual frontal air bags with a manual on-off switch for the RF passenger
 9 Unknown if equipped and/or what type

RDSUR – Road surface condition   Type: Numeric8
 0 Dry
 1 Wet, muddy, or oily
 2 Snow, ice, or slush
 9 Unknown

REGWTFA – Registration-year weight factor Type: Numeric8
 0 – 18437 Weight factor; share of the nation's vehicle registration years allocated to this induced-exposure case

ROLLCURT – Head curtain air bags designed to deploy in rollovers Type: Numeric8
 0 Not equipped with rollover curtain bags
 .01-.99 proportion of vehicles with rollover curtain bags (optional but cannot be identified from first 12 digits of VIN)
 1 Rollover curtains for driver and RF passenger (Note: CURTAIN should also be coded 1)

RURAL – County population density　　　　Type: Numeric8
    0　　　　Crash occurred in a county with ≥ 250 inhabitants per square mile in the 2000 census
    1　　　　Crash occurred in a county with < 250 inhabitants per square mile

SPDLIM55 – Speed limit 55+　　　　Type: Numeric8
    0　　　　Speed limit < 55 mph or not reported, for all roadways involved in the crash
    1　　　　Speed limit ≥ 55 mph for at least one roadway involved in the crash

SQUADCAR – Police model　　　　Type: Numeric8
    0　　　　Not a Ford Crown Victoria or Chevrolet Impala "police" model and not a 2004-2005 Chevrolet Impala SS
    1　　　　Ford Crown Victoria or Chevrolet Impala "police" model or 2004-2005 Chevrolet Impala SS

STATE – State where the crash occurred　　　　Type: Numeric8
    1　　　　Alabama
    12　　　Florida
    20　　　Kansas
    21　　　Kentucky
    24　　　Maryland
    26　　　Michigan
    29　　　Missouri
    34　　　New Jersey
    42　　　Pennsylvania
    53　　　Washington State
    55　　　Wisconsin
    56　　　Wyoming

TORSO – Side air bag with torso protection　　　　Type: Numeric8
    0　　　　Not equipped with torso bags
    .01-.99　proportion of vehicles with torso bags (optional but cannot be identified from first 12 digits of VIN)
    .25　　　Torso bag for RF passenger only
    1　　　　Torso bags for driver and RF passenger

TRAKWDTH – Track width (inches, average of front and rear wheels)　　Type: Numeric8
    54 – 72.65　Average of front and rear track width, in inches

TW_F – Track width, front wheels (inches)　　　　Type: Numeric8
    54.5 – 71.6　Front track width, in inches

TW_R – Track width, rear wheels (inches)　　　　Type: Numeric8
    52.2 – 75.8　Rear track width, in inches

V1 – 1st character of the VIN Type: Character1
V2 – 2nd character of the VIN Type: Character1
V3 – 3rd character of the VIN Type: Character1
V4 – 4th character of the VIN Type: Character1
V5 – 5th character of the VIN Type: Character1
V6 – 6th character of the VIN Type: Character1
V7 – 7th character of the VIN Type: Character1
V8 – 8th character of the VIN Type: Character1
V11 – 11th character of the VIN Type: Character1
V12 – 12th character of the VIN Type: Character1

VEHAGE – Vehicle age in years (CY – MY) Type: Numeric8
    0 – 8   Age of the case vehicle (years)

VMTWTFA – VMT weight factor Type: Numeric8
0 – 223,502,857 Weight factor; share of the nation's VMT allocated to this induced-exposure case

VTYP – Case vehicle type Type: Numeric8
    1   Passenger car, 2 doors
    2   Passenger car, 4 doors
    3   Pickup truck, light duty (compact or 150-series)
    4   Pickup truck, heavy duty (250- or 350-series)
    5   SUV, truck-based
    6   CUV (crossover SUV)
    7   Minivan, except Chevrolet Astro or GMC Safari
    7.1   Chevrolet Astro or GMC Safari
    8   Full-sized van

WB_MAX – Maximum wheelbase (inches, for selected pickup trucks) Type: Numeric8
    117.5 – 172.4 Maximum wheelbase for pickup trucks that can have 2 or more bed lengths for the same 12-character VIN (left blank for all other vehicles and not used in computing footprint)

WB_MIN – Wheelbase (inches) Type: Numeric8
    86.6 – 170.3 Wheelbase (and if this is a pickup truck, wheelbase for the shortest bed available for this 12-character VIN)

Appendix D: Volpe and Regression Coefficients for Sensitivity Tests

Baseline Model

	REGRESSION COEFFICIENTS						
	CURB WEIGHT		FOOTPRINT		FATALITIES AFTER ESC	N	FATALITY INCREASE PER 100-POUND MASS REDUCTION %
CRASH TYPE	COEFF	WALD CHI^2	COEFF	WALD CHI^2			
			CARS < 3,106 POUNDS				
1st-EVENT ROLLOVER	.0183	2.21	-.0808	31.56	207	-4	-1.83
HIT FIXED OBJECT	.0046	.46	-.0401	26.07	813	-4	-.46
HIT PEDESTRIAN/BIKE/MOTORCYCLE	-.0203	5.75	-.00914	.92	871	18	2.03
HIT HEAVY VEHICLE	-.0226	3.77	-.0297	4.86	471	11	2.26
HIT CAR-CUV-MINIVAN < 3082	-.00758	.57	-.00226	.04	478	4	.76
HIT CAR-CUV-MINIVAN 3082+	-.00478	.25	-.00489	.20	674	3	.48
HIT TRUCK-BASED LTV < 4150	-.0117	.98	-.0396	8.30	351	4	1.17
HIT TRUCK-BASED LTV 4150+	-.0606	29.80	-.0177	1.89	505	31	6.06
ALL OTHERS	-.0195	9.96	-.0114	2.64	1,530	30	1.95
VOLPE COEFFICIENT					5,901	92	1.56
			CARS ≥ 3,106 POUNDS				
1st-EVENT ROLLOVER	.0289	3.09	-.0808	31.56	247	-7	-2.89
HIT FIXED OBJECT	.0129	2.34	-.0401	26.07	1,263	-16	-1.29
HIT PEDESTRIAN/BIKE/MOTORCYCLE	.00141	.02	-.00914	.92	1,388	-2	-.14
HIT HEAVY VEHICLE	-.00391	.08	-.0297	4.86	687	3	.39
HIT CAR-CUV-MINIVAN < 3082	-.00261	.05	-.00226	.04	921	2	.26
HIT CAR-CUV-MINIVAN 3082+	-.0162	2.09	-.00489	.20	1,172	19	1.62
HIT TRUCK-BASED LTV < 4150	-.00531	.14	-.0396	8.30	543	3	.53
HIT TRUCK-BASED LTV 4150+	-.0234	3.11	-.0177	1.89	727	17	2.34
ALL OTHERS	-.0116	2.51	-.0114	2.64	2,552	30	1.16
VOLPE COEFFICIENT					9,499	48	.51

Baseline Model

	REGRESSION COEFFICIENTS					FATALITY INCREASE PER 100-POUND MASS REDUCTION	
	CURB WEIGHT		FOOTPRINT		FATALITIES AFTER ESC		
CRASH TYPE	COEFF	WALD CHI^2	COEFF	WALD CHI^2		N	%
	PICKUPS & TRUCK-BASED SUVs < 4,594						
1st-EVENT ROLLOVER	-.00655	1.40	-.0119	7.65	162	1	.66
HIT FIXED OBJECT	.0139	7.03	-.0199	26.73	451	-6	-1.39
HIT PEDESTRIAN/BIKE/MOTORCYCLE	-.0107	3.41	.0124	8.60	676	7	1.07
HIT HEAVY VEHICLE	-.0162	3.60	-.0075	1.41	287	5	1.62
HIT CAR-CUV-MINIVAN < 3082	.00091	.02	.00207	.24	530	-0	-.09
HIT CAR-CUV-MINIVAN 3082+	.00708	1.30	-.00308	.46	485	-3	-.71
HIT TRUCK-BASED LTV < 4150	.00625	.56	-.0101	2.81	252	-2	-.63
HIT TRUCK-BASED LTV 4150+	-.0446	26.58	.0169	6.88	289	13	4.46
ALL OTHERS	-.00733	2.86	.00442	1.93	1,126	8	.73
VOLPE COEFFICIENT					4,258	22	.52
	PICKUPS & TRUCK-BASED SUVs ≥ 4,594						
1st-EVENT ROLLOVER	.0128	7.51	-.0119	7.65	345	-4	-1.28
HIT FIXED OBJECT	-.00763	2.74	-.0199	26.73	802	6	.76
HIT PEDESTRIAN/BIKE/MOTORCYCLE	.00055	.01	.0124	8.60	1,516	-1	-.06
HIT HEAVY VEHICLE	-.0032	.17	-.0075	1.41	492	2	.32
HIT CAR-CUV-MINIVAN < 3082	.0091	3.99	.00207	.24	1,262	-12	-.91
HIT CAR-CUV-MINIVAN 3082+	.0137	8.05	-.00308	.46	1,155	-16	-1.37
HIT TRUCK-BASED LTV < 4150	.00962	1.99	-.0101	2.81	578	-6	-.96
HIT TRUCK-BASED LTV 4150+	-.00534	.53	.0169	6.88	490	3	.53
ALL OTHERS	.00112	.10	.00442	1.93	2,262	-3	-.11
VOLPE COEFFICIENT					8,902	-30	-.34
	CUVs & MINIVANS						
1st-EVENT ROLLOVER	.0702	15.31	-.1159	22.66	100	-7	-7.02
HIT FIXED OBJECT	.0361	7.47	-.0767	18.19	373	-13	-3.61
HIT PEDESTRIAN/BIKE/MOTORCYCLE	.0157	2.19	-.00371	.07	812	-13	-1.57
HIT HEAVY VEHICLE	-.0194	1.10	-.0466	3.76	297	6	1.94
HIT CAR-CUV-MINIVAN < 3082	.0009	.01	.00787	.19	503	-0	-.09
HIT CAR-CUV-MINIVAN 3082+	-.0168	1.54	.0219	1.47	569	10	1.68
HIT TRUCK-BASED LTV < 4150	-.0382	3.94	.0405	2.48	244	9	3.82
HIT TRUCK-BASED LTV 4150+	.00927	.29	-.0380	2.64	294	-3	-.93
ALL OTHERS	.00396	.23	-.0272	5.98	1,380	-5	-.40
VOLPE COEFFICIENT					4,571	-17	-.37

Test No. 1: By Track Width and Wheelbase, With Stopped-Vehicle State Data

	REGRESSION COEFFICIENTS							FATALITY INCREASE PER 100-LB MASS REDUCTION	
	CURB WEIGHT		TRACK WIDTH		WHEELBASE		FATALITIES AFTER ESC		
CRASH TYPE	COEFF	WALD CHI^2	COEFF	WALD CHI^2	COEFF	WALD CHI^2		N	%
			CARS < 3,106 POUNDS						
1st-EVENT ROLLOVER	.0416	10.10	-.1818	61.04	-.0192	6.03	207	-9	-4.16
HIT FIXED OBJECT	.0235	10.79	-.1085	74.74	-.0103	5.29	813	-19	-2.35
HIT PEDESTRIAN/BIKE/MOTORCYCLE	-.0141	2.51	-.0295	3.57	-.0039	.52	871	12	1.41
HIT HEAVY VEHICLE	-.00719	.35	-.0689	10.56	-.0120	2.43	471	3	.72
HIT CAR-CUV-MINIVAN < 3082	-.00004	.00	-.0297	2.61	-.00186	.08	478	0	.00
HIT CAR-CUV-MINIVAN 3082+	.00103	1.09	-.0646	14.08	.00528	.69	674	-7	-1.03
HIT TRUCK-BASED LTV < 4150	.00503	.16	-.0825	14.18	-.0112	2.03	351	-2	-.50
HIT TRUCK-BASED LTV 4150+	-.0339	8.56	-.1225	36.52	.0124	2.81	505	17	3.39
ALL OTHERS	-.0124	3.67	-.0330	8.74	-.00441	1.20	1,530	19	1.24
VOLPE COEFFICIENT							5,901	15	.25
			CARS ≥ 3,106 POUNDS						
1st-EVENT ROLLOVER	.0586	12.62	-.1818	61.04	-.0192	6.03	247	-14	-5.86
HIT FIXED OBJECT	.0389	21.55	-.1085	74.74	-.0103	5.29	1,263	-49	-3.89
HIT PEDESTRIAN/BIKE/MOTORCYCLE	.00889	.82	-.0295	3.57	-.0039	.52	1,388	-12	-.89
HIT HEAVY VEHICLE	.0167	1.49	-.0689	10.56	-.0120	2.43	687	-11	-1.67
HIT CAR-CUV-MINIVAN < 3082	.00754	.41	-.0297	2.61	-.00186	.08	921	-7	-.75
HIT CAR-CUV-MINIVAN 3082+	.0010	.01	-.0646	14.08	.00528	.69	1,172	-1	-.10
HIT TRUCK-BASED LTV < 4150	.00707	.25	-.0825	14.18	-.0112	2.03	543	-4	-.71
HIT TRUCK-BASED LTV 4150+	.0021	.03	-.1225	36.52	.0124	2.81	727	-5	-.21
ALL OTHERS	-.00619	.73	-.0330	8.74	-.00441	1.20	2,552	16	.62
VOLPE COEFFICIENT							9,499	-85	-.89

147

Test No. 1: By Track Width and Wheelbase, With Stopped-Vehicle State Data

REGRESSION COEFFICIENTS

CRASH TYPE	CURB WEIGHT COEFF	WALD CHI^2	TRACK WIDTH COEFF	WALD CHI^2	WHEELBASE COEFF	WALD CHI^2	FATALITIES AFTER ESC	FATALITY INCREASE PER 100-LB MASS REDUCTION N	%
PICKUPS & TRUCK-BASED SUVs < 4,594								
1st-EVENT ROLLOVER	.0262	15.13	-.0710	86.16	.0007	.12	162	-4	-2.62
HIT FIXED OBJECT	.0122	3.76	-.00758	1.13	-.0101	31.60	451	-6	-1.22
HIT PEDESTRIAN/BIKE/MOTORCYCLE	-.0125	3.13	.0144	3.15	.00492	6.19	676	8	1.25
HIT HEAVY VEHICLE	-.0132	1.64	-.0155	1.76	-.00241	.68	287	4	1.32
HIT CAR-CUV-MINIVAN < 3082	.0144	4.07	-.0185	5.16	-.00182	.84	530	-8	-1.44
HIT CAR-CUV-MINIVAN 3082+	.0222	8.30	-.0266	9.21	-.00037	.03	485	-11	-2.22
HIT TRUCK-BASED LTV < 4150	.00785	.10	-.00894	.59	-.00469	2.83	252	-2	-.79
HIT TRUCK-BASED LTV 4150+	-.0272	6.63	-.0230	3.64	.0109	13.10	289	8	2.72
ALL OTHERS	-.00516	.97	.00262	.19	.00216	2.09	1,126	6	.52
VOLPE COEFFICIENT							4,258	-4	-.09
PICKUPS & TRUCK-BASED SUVs ≥ 4,594								
1st-EVENT ROLLOVER	.0307	40.15	-.0710	86.16	.0007	.12	345	-11	-3.07
HIT FIXED OBJECT	-.00243	.26	-.00758	1.13	-.0101	31.60	802	2	.24
HIT PEDESTRIAN/BIKE/MOTORCYCLE	.00361	.56	.0144	3.15	.00492	6.19	1,516	-5	-.36
HIT HEAVY VEHICLE	.00428	.30	-.0155	1.76	-.00241	.68	492	-2	-.43
HIT CAR-CUV-MINIVAN < 3082	.0170	12.57	-.0185	5.16	-.00182	.84	1,262	-21	-1.70
HIT CAR-CUV-MINIVAN 3082+	.0224	19.40	-.0266	9.21	-.00037	.03	1,155	-26	-2.24
HIT TRUCK-BASED LTV < 4150	.0149	4.43	-.00894	.59	-.00469	2.83	578	-9	-1.49
HIT TRUCK-BASED LTV 4150+	.00618	.67	-.0230	3.64	.0109	13.10	490	-3	-.62
ALL OTHERS	.00467	1.65	.00262	.19	.00216	2.09	2,262	-11	-.47
VOLPE COEFFICIENT							8,902	-86	-.97
CUVs & MINIVANS								
1st-EVENT ROLLOVER	.0863	22.17	-.1823	27.41	-.0240	2.89	100	-9	-8.63
HIT FIXED OBJECT	.0375	7.92	-.1025	16.62	-.0206	3.94	373	-14	-3.75
HIT PEDESTRIAN/BIKE/MOTORCYCLE	.00999	.84	.0278	2.08	-.0117	2.08	812	-8	-1.00
HIT HEAVY VEHICLE	-.0179	.94	-.0745	4.63	-.00788	.34	297	5	1.79
HIT CAR-CUV-MINIVAN < 3082	-.00081	.00	.0211	.76	-.0031	.09	503	-0	-.08
HIT CAR-CUV-MINIVAN 3082+	-.0157	1.32	.0235	.94	.00317	.10	569	9	1.57
HIT TRUCK-BASED LTV < 4150	-.0469	5.80	.0831	5.56	.00457	.10	244	11	4.69
HIT TRUCK-BASED LTV 4150+	.00606	.12	-.0214	.42	-.0173	1.73	294	-2	-.61
ALL OTHERS	-.0007	.01	.0313	4.41	-.0299	22.66	1,380	1	.07
VOLPE COEFFICIENT							4,571	-6	-.13

Test No. 2: With Stopped-Vehicle State Data

	REGRESSION COEFFICIENTS						FATALITY INCREASE PER 100-POUND MASS REDUCTION	
	CURB WEIGHT		FOOTPRINT					
CRASH TYPE	COEFF	WALD CHI^2	COEFF	WALD CHI^2		FATALITIES AFTER ESC	N	%
				CARS < 3,106 POUNDS				
1st-EVENT ROLLOVER	.0192	2.48	-.1092	59.30		207	-4	-1.92
HIT FIXED OBJECT	.0113	2.79	-.0654	69.55		813	-9	-1.13
HIT PEDESTRIAN/BIKE/MOTORCYCLE	-.0174	4.24	-.0188	3.88		871	15	1.74
HIT HEAVY VEHICLE	-.0132	1.27	-.0505	13.85		471	6	1.32
HIT CAR-CUV-MINIVAN < 3082	-.00304	.09	-.0169	2.12		478	1	.30
HIT CAR-CUV-MINIVAN 3082+	.00187	.04	-.0210	3.64		674	-1	-.19
HIT TRUCK-BASED LTV < 4150	-.00302	.06	-.0551	15.89		351	1	.30
HIT TRUCK-BASED LTV 4150+	-.0498	20.11	-.0366	7.98		505	25	4.98
ALL OTHERS	-.0149	5.75	-.0231	10.76		1,530	23	1.49
VOLPE COEFFICIENT						5,901	57	.97
				CARS ≥ 3,106 POUNDS				
1st-EVENT ROLLOVER	.0470	8.10	-.1092	59.30		247	-12	-4.70
HIT FIXED OBJECT	.0349	16.95	-.0654	69.55		1,263	-44	-3.49
HIT PEDESTRIAN/BIKE/MOTORCYCLE	.00771	.60	-.0188	3.88		1,388	-11	-.77
HIT HEAVY VEHICLE	.0143	1.07	-.0505	13.85		687	-10	-1.43
HIT CAR-CUV-MINIVAN < 3082	.00672	.31	-.0169	2.12		921	-6	-.67
HIT CAR-CUV-MINIVAN 3082+	-.00312	.01	-.0210	3.64		1,172	4	.31
HIT TRUCK-BASED LTV < 4150	.00325	.05	-.0551	15.89		543	-2	-.33
HIT TRUCK-BASED LTV 4150+	-.00746	.31	-.0366	7.98		727	5	.75
ALL OTHERS	-.0062	.71	-.0231	10.76		2,552	16	.62
VOLPE COEFFICIENT						9,499	-59	-.62

Test No. 2: With Stopped-Vehicle State Data

	REGRESSION COEFFICIENTS				FATALITIES AFTER ESC	FATALITY INCREASE PER 100-POUND MASS REDUCTION	
	CURB WEIGHT		FOOTPRINT				
CRASH TYPE	COEFF	WALD CHI²	COEFF	WALD CHI²		N	%
			PICKUPS & TRUCK-BASED SUVs < 4,594				
1st-EVENT ROLLOVER	-.00234	.17	-.0111	6.63	162	0	.23
HIT FIXED OBJECT	.0178	11.42	-.0216	31.90	451	-8	-1.78
HIT PEDESTRIAN/BIKE/MOTORCYCLE	-.0106	3.30	.0124	8.61	676	7	1.06
HIT HEAVY VEHICLE	-.0175	4.17	-.0078	1.56	287	5	1.75
HIT CAR-CUV-MINIVAN < 3082	.00427	.53	.00115	.07	530	-2	-.43
HIT CAR-CUV-MINIVAN 3082+	.0093	2.20	-.0037	.67	485	-5	-.93
HIT TRUCK-BASED LTV < 4150	.00765	.83	-.0106	3.14	252	-2	-.77
HIT TRUCK-BASED LTV 4150+	-.0446	26.36	.0169	6.91	289	13	4.46
ALL OTHERS	-.00559	1.65	.00455	2.03	1,126	6	.56
VOLPE COEFFICIENT					4,258	15	.35
			PICKUPS & TRUCK-BASED SUVs ≥ 4,594				
1st-EVENT ROLLOVER	.0202	18.20	-.0111	6.63	345	-7	-2.02
HIT FIXED OBJECT	.00003	.00	-.0216	31.90	802	0	.00
HIT PEDESTRIAN/BIKE/MOTORCYCLE	.00393	.72	.0124	8.61	1,516	-6	-.39
HIT HEAVY VEHICLE	.0033	.18	-.0078	1.56	492	-2	-.33
HIT CAR-CUV-MINIVAN < 3082	.0133	8.36	.00115	.07	1,262	-17	-1.33
HIT CAR-CUV-MINIVAN 3082+	.0178	13.38	-.0037	.67	1,155	-21	-1.78
HIT TRUCK-BASED LTV < 4150	.0151	4.86	-.0106	3.14	578	-9	-1.51
HIT TRUCK-BASED LTV 4150+	.00039	.00	.0169	6.91	490	-0	-.04
ALL OTHERS	.00443	1.61	.00455	2.03	2,262	-10	-.44
VOLPE COEFFICIENT					8,902	-71	-.80
			CUVs & MINIVANS				
1st-EVENT ROLLOVER	.0737	17.29	-.1208	24.05	100	-7	-7.37
HIT FIXED OBJECT	.0325	6.17	-.0789	19.20	373	-12	-3.25
HIT PEDESTRIAN/BIKE/MOTORCYCLE	.0145	1.86	-.00551	.14	812	-12	-1.45
HIT HEAVY VEHICLE	-.0224	1.53	-.0459	3.69	297	7	2.24
HIT CAR-CUV-MINIVAN < 3082	.00294	.05	.00533	.09	503	-1	-.29
HIT CAR-CUV-MINIVAN 3082+	-.0154	1.34	.0179	.99	569	9	1.54
HIT TRUCK-BASED LTV < 4150	-.0393	4.31	.0428	2.80	244	10	3.93
HIT TRUCK-BASED LTV 4150+	.00583	.12	-.0344	2.16	294	-2	-.58
ALL OTHERS	.00395	.23	-.0271	5.99	1,380	-5	-.40
VOLPE COEFFICIENT					4,571	-15	-.33

Test No. 3: By Track Width and Wheelbase

CRASH TYPE	CURB WEIGHT		TRACK WIDTH		WHEELBASE		FATALITIES AFTER ESC	FATALITY INCREASE PER 100-LB MASS REDUCTION	
	COEFF	WALD CHI^2	COEFF	WALD CHI^2	COEFF	WALD CHI^2		N	%
			CARS < 3,106 POUNDS						
1st-EVENT ROLLOVER	.0394	8.85	-.1528	43.13	-.0104	1.78	207	-8	-3.94
HIT FIXED OBJECT	.0151	4.45	-.0824	43.22	-.00214	.23	813	-12	-1.51
HIT PEDESTRIAN/BIKE/MOTORCYCLE	-.0171	3.67	-.0223	2.05	-.00004	.00	871	15	1.71
HIT HEAVY VEHICLE	-.0188	2.40	-.0400	3.57	-.00763	1.01	471	9	1.88
HIT CAR-CUV-MINIVAN < 3082	-.00576	.30	-.0124	.46	.0022	.11	478	3	.58
HIT CAR-CUV-MINIVAN 3082+	.00228	.05	-.0443	6.66	.00931	2.22	674	-2	-.23
HIT TRUCK-BASED LTV < 4150	-.00506	.17	-.0627	8.26	-.00744	.91	351	2	.51
HIT TRUCK-BASED LTV 4150+	-.0466	16.18	-.0957	22.55	.0162	4.88	505	24	4.66
ALL OTHERS	-.0175	7.32	-.0219	3.83	-.00042	.01	1,530	27	1.75
VOLPE COEFFICIENT							5,901	57	.97
			CARS ≥ 3,106 POUNDS						
1st-EVENT ROLLOVER	.0395	5.80	-.1528	43.13	-.0104	1.78	247	-10	-3.95
HIT FIXED OBJECT	.0169	4.09	-.0824	43.22	-.00214	.23	1,263	-21	-1.69
HIT PEDESTRIAN/BIKE/MOTORCYCLE	.00282	.08	-.0223	2.05	-.00004	.00	1,388	-4	-.28
HIT HEAVY VEHICLE	-.00211	.02	-.0400	3.57	-.00763	1.01	687	1	.21
HIT CAR-CUV-MINIVAN < 3082	-.00183	.02	-.0124	.46	.0022	.11	921	2	.18
HIT CAR-CUV-MINIVAN 3082+	.0122	1.22	-.0443	6.66	.00931	2.22	1,172	14	1.22
HIT TRUCK-BASED LTV < 4150	-.00181	.02	-.0627	8.26	-.00744	.91	543	1	.18
HIT TRUCK-BASED LTV 4150+	-.0142	1.17	-.0957	22.55	.0162	4.88	727	10	1.42
ALL OTHERS	-.0114	2.46	-.0219	3.83	-.00042	.01	2,552	29	1.14
VOLPE COEFFICIENT							9,499	23	.24

151

Test No. 3: By Track Width and Wheelbase

	REGRESSION COEFFICIENTS						FATALITIES AFTER ESC	FATALITY INCREASE PER 100-LB MASS REDUCTION	
	CURB WEIGHT		TRACK WIDTH		WHEELBASE				
CRASH TYPE	COEFF	WALD CHI^2	COEFF	WALD CHI^2	COEFF	WALD CHI^2		N	%
			PICKUPS & TRUCK-BASED SUVs < 4,594						
1st-EVENT ROLLOVER	.0222	11.14	-.0683	81.35	-.00076	.14	162	-4	-2.22
HIT FIXED OBJECT	.00756	1.45	-.00411	.33	-.00972	29.33	451	-3	-.76
HIT PEDESTRIAN/BIKE/MOTORCYCLE	-.0111	2.50	.0117	2.08	.00508	6.59	676	8	1.11
HIT HEAVY VEHICLE	-.0110	1.12	-.0155	1.75	-.00277	.89	287	3	1.10
HIT CAR-CUV-MINIVAN < 3082	.0125	3.05	-.0199	5.92	.00215	1.19	530	-7	-1.25
HIT CAR-CUV-MINIVAN 3082+	.0224	8.41	-.0298	11.53	-.00013	.00	485	-11	-2.24
HIT TRUCK-BASED LTV < 4150	.00991	.92	-.0140	1.45	-.0044	2.50	252	-2	-.99
HIT TRUCK-BASED LTV 4150+	-.0256	5.82	-.0238	3.88	.0104	11.88	289	7	2.56
ALL OTHERS	-.00541	1.06	.00003	.00	.00214	2.08	1,126	6	.54
VOLPE COEFFICIENT							4,258	-3	-.07
			PICKUPS & TRUCK-BASED SUVs ≥ 4,594						
1st-EVENT ROLLOVER	.0241	25.14	-.0683	81.35	-.00076	.14	345	-8	-2.41
HIT FIXED OBJECT	-.0103	4.76	-.00411	.33	-.00972	29.33	802	8	1.03
HIT PEDESTRIAN/BIKE/MOTORCYCLE	.00079	.03	.0117	2.08	.00508	6.59	1,516	-1	-.08
HIT HEAVY VEHICLE	-.00154	.04	-.0155	1.75	-.00277	.89	492	1	.15
HIT CAR-CUV-MINIVAN < 3082	.0136	8.05	-.0199	5.92	.00215	1.19	1,262	-17	-1.36
HIT CAR-CUV-MINIVAN 3082+	.0194	14.64	-.0298	11.53	-.00013	.00	1,155	-22	-1.94
HIT TRUCK-BASED LTV < 4150	.0107	2.30	-.0140	1.45	-.0044	2.50	578	-6	-1.07
HIT TRUCK-BASED LTV 4150+	.00162	.05	-.0238	3.88	.0104	11.88	490	-1	-.16
ALL OTHERS	.00199	.30	.00003	.00	.00214	2.08	2,262	-5	-.20
VOLPE COEFFICIENT							8,902	-52	-.58
			CUVs & MINIVANS						
1st-EVENT ROLLOVER	.0792	18.47	-.1580	22.75	-.0262	3.54	100	-8	-7.92
HIT FIXED OBJECT	.0416	9.60	-.1068	19.06	-.0171	2.80	373	-16	-4.16
HIT PEDESTRIAN/BIKE/MOTORCYCLE	.0112	1.06	.0284	2.21	-.0107	1.74	812	-9	-1.12
HIT HEAVY VEHICLE	-.0137	.53	-.0886	6.93	-.00489	.08	297	4	1.37
HIT CAR-CUV-MINIVAN < 3082	-.00075	.00	.0178	.55	-.00014	.00	503	0	.08
HIT CAR-CUV-MINIVAN 3082+	-.0164	1.41	.0201	.71	.00709	.50	569	9	1.64
HIT TRUCK-BASED LTV < 4150	-.0440	4.95	.0667	3.70	.00798	.32	244	11	4.40
HIT TRUCK-BASED LTV 4150+	.0114	.41	-.0415	1.68	-.0133	1.05	294	-3	-1.14
ALL OTHERS	-.00001	.00	.0223	2.27	-.0267	18.23	1,380	0	.00
VOLPE COEFFICIENT							4,571	-11	-.24

Test No. 4: Without CY Control Variables

	REGRESSION COEFFICIENTS						FATALITY INCREASE PER 100-POUND MASS REDUCTION	
	CURB WEIGHT		FOOTPRINT					
CRASH TYPE	COEFF	WALD CHI^2	COEFF	WALD CHI^2	FATALITIES AFTER ESC		N	%
			CARS < 3,106 POUNDS					
1st-EVENT ROLLOVER	.0169	1.87	-.0818	32.30	207		-4	-1.69
HIT FIXED OBJECT	.00461	.46	-.0403	26.52	813		-4	-.46
HIT PEDESTRIAN/BIKE/MOTORCYCLE	-.0200	5.61	-.00788	.69	871		17	2.00
HIT HEAVY VEHICLE	-.0222	3.63	-.0310	5.32	471		10	2.22
HIT CAR-CUV-MINIVAN < 3082	-.00656	.42	-.0133	1.36	478		3	.66
HIT CAR-CUV-MINIVAN 3082+	-.00444	.22	-.00387	.13	674		3	.44
HIT TRUCK-BASED LTV < 4150	-.0126	1.12	-.0482	12.32	351		4	1.26
HIT TRUCK-BASED LTV 4150+	-.0600	29.28	-.0181	1.98	505		30	6.00
ALL OTHERS	-.0188	9.26	-.0152	4.74	1,530		29	1.88
VOLPE COEFFICIENT					**5,901**		**90**	**1.53**
			CARS ≥ 3,106 POUNDS					
1st-EVENT ROLLOVER	.0294	3.21	-.0818	32.30	247		-7	-2.94
HIT FIXED OBJECT	.0129	2.34	-.0403	26.52	1,263		-16	-1.29
HIT PEDESTRIAN/BIKE/MOTORCYCLE	.00056	.00	-.00788	.69	1,388		-1	-.06
HIT HEAVY VEHICLE	-.0037	.07	-.0310	5.32	687		3	.37
HIT CAR-CUV-MINIVAN < 3082	-.00077	.00	-.0133	1.36	921		-1	-.08
HIT CAR-CUV-MINIVAN 3082+	-.0163	2.12	-.00387	.13	1,172		19	1.63
HIT TRUCK-BASED LTV < 4150	-.00348	.06	-.0482	12.32	543		2	.35
HIT TRUCK-BASED LTV 4150+	-.0233	3.06	-.0181	1.98	727		17	2.33
ALL OTHERS	-.0101	1.90	-.0152	4.74	2,552		26	1.01
VOLPE COEFFICIENT					**9,499**		**41**	**.43**

Test No. 4: Without CY Control Variables

	REGRESSION COEFFICIENTS					FATALITY INCREASE PER 100-POUND MASS REDUCTION	
	CURB WEIGHT		FOOTPRINT		FATALITIES AFTER ESC		
CRASH TYPE	COEFF	WALD CHI^2	COEFF	WALD CHI^2		N	%
			PICKUPS & TRUCK-BASED SUVs < 4,594				
1st-EVENT ROLLOVER	-.0124	5.21	-.00911	4.56	162	2	1.24
HIT FIXED OBJECT	.0111	4.64	-.0185	23.41	451	-5	-1.11
HIT PEDESTRIAN/BIKE/MOTORCYCLE	-.0114	4.02	.0128	9.24	676	8	1.14
HIT HEAVY VEHICLE	-.0207	6.13	-.00513	.67	287	6	2.07
HIT CAR-CUV-MINIVAN < 3082	-.0142	6.29	.00974	5.29	530	8	1.42
HIT CAR-CUV-MINIVAN 3082+	-.00049	.01	.00027	.00	485	-0	-.05
HIT TRUCK-BASED LTV < 4150	-.00971	1.44	-.0017	.08	252	2	.97
HIT TRUCK-BASED LTV 4150+	-.0463	29.61	.0178	7.76	289	13	4.63
ALL OTHERS	-.0157	13.74	.00854	7.28	1,126	18	1.57
VOLPE COEFFICIENT					4,258	41	1.20
			PICKUPS & TRUCK-BASED SUVs ≥ 4,594				
1st-EVENT ROLLOVER	.00813	3.04	-.00911	4.56	345	-3	-.81
HIT FIXED OBJECT	-.0104	5.29	-.0185	23.41	802	8	1.04
HIT PEDESTRIAN/BIKE/MOTORCYCLE	-.00007	.00	.0128	9.24	1,516	0	.01
HIT HEAVY VEHICLE	-.00777	1.05	-.00513	.67	492	4	.78
HIT CAR-CUV-MINIVAN < 3082	-.0045	.95	.00974	5.29	1,262	6	-.45
HIT CAR-CUV-MINIVAN 3082+	.00788	2.70	.00027	.00	1,155	-9	-.79
HIT TRUCK-BASED LTV < 4150	-.00636	.86	-.0017	.08	578	4	.64
HIT TRUCK-BASED LTV 4150+	-.00701	.95	.0178	7.76	490	3	.70
ALL OTHERS	-.00613	3.12	.00854	7.28	2,262	14	.61
VOLPE COEFFICIENT					8,902	27	.30
			CUVs & MINIVANS				
1st-EVENT ROLLOVER	.0642	12.83	-.1180	23.33	100	-6	-6.42
HIT FIXED OBJECT	.0363	7.68	-.0769	18.29	373	-14	-3.63
HIT PEDESTRIAN/BIKE/MOTORCYCLE	.0160	2.27	-.00316	.05	812	-13	-1.60
HIT HEAVY VEHICLE	-.0194	1.14	-.0470	3.83	297	6	1.94
HIT CAR-CUV-MINIVAN < 3082	-.00981	.55	.00965	.28	503	5	.98
HIT CAR-CUV-MINIVAN 3082+	-.0186	1.92	.0225	1.56	569	11	1.86
HIT TRUCK-BASED LTV < 4150	-.0580	8.91	.0489	3.54	244	14	5.80
HIT TRUCK-BASED LTV 4150+	.00493	.08	-.0369	2.48	294	-1	-.49
ALL OTHERS	-.00035	.00	-.0262	5.51	1,380	0	.04
VOLPE COEFFICIENT					4,571	2	.04

Test No. 5: CUVs and Minivans Weighted According to 2010 Sales

	REGRESSION COEFFICIENTS					FATALITY INCREASE PER 100-POUND MASS REDUCTION	
	CURB WEIGHT		FOOTPRINT		FATALITIES AFTER ESC		
CRASH TYPE	COEFF	WALD CHI^2	COEFF	WALD CHI^2		N	%
			CARS < 3,106 POUNDS				
1st-EVENT ROLLOVER	.0183	2.21	-.0808	31.56	207	-4	-1.83
HIT FIXED OBJECT	.0046	.46	-.0401	26.07	813	-4	-.46
HIT PEDESTRIAN/BIKE/MOTORCYCLE	-.0203	5.75	-.00914	.92	871	18	2.03
HIT HEAVY VEHICLE	-.0226	3.77	-.0297	4.86	471	11	2.26
HIT CAR-CUV-MINIVAN < 3082	-.00758	.57	-.00226	.04	478	4	.76
HIT CAR-CUV-MINIVAN 3082+	-.00478	.25	-.00489	.20	674	3	.48
HIT TRUCK-BASED LTV < 4150	-.0117	.98	-.0396	8.30	351	4	1.17
HIT TRUCK-BASED LTV 4150+	-.0606	29.80	-.0177	1.89	505	31	6.06
ALL OTHERS	-.0195	9.96	-.0114	2.64	1,530	30	1.95
VOLPE COEFFICIENT					5,901	92	1.56
			CARS ≥ 3,106 POUNDS				
1st-EVENT ROLLOVER	.0289	3.09	-.0808	31.56	247	-7	-2.89
HIT FIXED OBJECT	.0129	2.34	-.0401	26.07	1,263	-16	-1.29
HIT PEDESTRIAN/BIKE/MOTORCYCLE	.00141	.02	-.00914	.92	1,388	-2	-.14
HIT HEAVY VEHICLE	-.00391	.08	-.0297	4.86	687	3	.39
HIT CAR-CUV-MINIVAN < 3082	-.00261	.05	-.00226	.04	921	2	.26
HIT CAR-CUV-MINIVAN 3082+	-.0162	2.09	-.00489	.20	1,172	19	1.62
HIT TRUCK-BASED LTV < 4150	-.00531	.14	-.0396	8.30	543	3	.53
HIT TRUCK-BASED LTV 4150+	-.0234	3.11	-.0177	1.89	727	17	2.34
ALL OTHERS	-.0116	2.51	-.0114	2.64	2,552	30	1.16
VOLPE COEFFICIENT					9,499	48	.51

Test No. 5: CUVs and Minivans Weighted According to 2010 Sales

	REGRESSION COEFFICIENTS				FATALITIES AFTER ESC	FATALITY INCREASE PER 100-POUND MASS REDUCTION	
	CURB WEIGHT		FOOTPRINT				
CRASH TYPE	COEFF	WALD CHI²	COEFF	WALD CHI²		N	%
	PICKUPS & TRUCK-BASED SUVs < 4,594						
1st-EVENT ROLLOVER	-.00655	1.40	-.0119	7.65	162	1	.66
HIT FIXED OBJECT	.0139	7.03	-.0199	26.73	451	-6	-1.39
HIT PEDESTRIAN/BIKE/MOTORCYCLE	-.0107	3.41	.0124	8.60	676	7	1.07
HIT HEAVY VEHICLE	-.0162	3.60	-.0075	1.41	287	5	1.62
HIT CAR-CUV-MINIVAN < 3082	.00091	.02	-.00207	.24	530	-0	-.09
HIT CAR-CUV-MINIVAN 3082+	-.00708	1.30	-.00308	.46	485	-3	-.71
HIT TRUCK-BASED LTV < 4150	-.00625	.56	-.0101	2.81	252	-2	-.63
HIT TRUCK-BASED LTV 4150+	-.0446	26.58	.0169	6.88	289	13	4.46
ALL OTHERS	-.00733	2.86	.00442	1.93	1,126	8	.73
VOLPE COEFFICIENT					4,258	22	.52
	PICKUPS & TRUCK-BASED SUVs ≥ 4,594						
1st-EVENT ROLLOVER	.0128	7.51	-.0119	7.65	345	-4	-1.28
HIT FIXED OBJECT	-.00763	2.74	-.0199	26.73	802	6	.76
HIT PEDESTRIAN/BIKE/MOTORCYCLE	.00055	.01	.0124	8.60	1,516	-1	-.06
HIT HEAVY VEHICLE	-.0032	.17	-.0075	1.41	492	2	.32
HIT CAR-CUV-MINIVAN < 3082	.0091	3.99	-.00207	.24	1,262	-12	-.91
HIT CAR-CUV-MINIVAN 3082+	.0137	8.05	-.00308	.46	1,155	-16	-1.37
HIT TRUCK-BASED LTV < 4150	-.00962	1.99	-.0101	2.81	578	-6	-.96
HIT TRUCK-BASED LTV 4150+	-.00534	.53	.0169	6.88	490	3	.53
ALL OTHERS	.00112	.10	.00442	1.93	2,262	-3	-.11
VOLPE COEFFICIENT					8,902	-30	-.34
	CUVs & MINIVANS						
1st-EVENT ROLLOVER	.0648	11.55	-.0998	14.81	105	-7	-6.48
HIT FIXED OBJECT	.0164	1.28	-.0531	7.59	413	-7	-1.64
HIT PEDESTRIAN/BIKE/MOTORCYCLE	.00457	.17	.0128	.70	836	-4	-.46
HIT HEAVY VEHICLE	-.0165	.62	-.0496	3.28	276	5	1.65
HIT CAR-CUV-MINIVAN < 3082	-.0278	3.69	.0439	5.13	520	14	2.78
HIT CAR-CUV-MINIVAN 3082+	-.00767	.27	.0129	.44	529	4	.77
HIT TRUCK-BASED LTV < 4150	-.0468	4.84	.0608	4.62	224	10	4.68
HIT TRUCK-BASED LTV 4150+	-.0109	.30	-.0153	.33	250	3	1.09
ALL OTHERS	-.00376	.17	-.0235	3.77	1,353	5	.38
VOLPE COEFFICIENT					4,507	24	.53

Test No. 6: Without Non-Significant Control Variables

REGRESSION COEFFICIENTS

CRASH TYPE	CURB WEIGHT		FOOTPRINT		FATALITIES AFTER ESC	FATALITY INCREASE PER 100-POUND MASS REDUCTION	
	COEFF	WALD CHI²	COEFF	WALD CHI²		N	%
			CARS < 3,106 POUNDS				
1st-EVENT ROLLOVER	.0177	2.07	-.0816	32.28	207	-4	-1.77
HIT FIXED OBJECT	.00378	.31	-.0402	26.33	813	-3	-.38
HIT PEDESTRIAN/BIKE/MOTORCYCLE	-.0210	6.28	-.00761	.70	871	18	2.10
HIT HEAVY VEHICLE	-.0228	3.94	-.0282	4.65	471	11	2.28
HIT CAR-CUV-MINIVAN < 3082	-.00959	.93	-.00007	.00	478	5	.96
HIT CAR-CUV-MINIVAN 3082+	-.00675	.52	-.00138	.02	674	5	.68
HIT TRUCK-BASED LTV < 4150	-.0126	1.13	-.0397	8.62	351	4	1.26
HIT TRUCK-BASED LTV 4150+	-.0612	32.91	-.0158	1.55	505	31	6.12
ALL OTHERS	-.0196	10.55	-.0102	2.45	1,530	30	1.96
VOLPE COEFFICIENT					5,901	97	1.64
			CARS ≥ 3,106 POUNDS				
1st-EVENT ROLLOVER	.0305	3.48	-.0816	32.28	247	-8	-3.05
HIT FIXED OBJECT	.0135	2.56	-.0402	26.33	1,263	-17	-1.35
HIT PEDESTRIAN/BIKE/MOTORCYCLE	-.00172	.04	-.00761	.70	1,388	2	.17
HIT HEAVY VEHICLE	-.0082	.38	-.0282	4.65	687	6	-.82
HIT CAR-CUV-MINIVAN < 3082	-.00552	.25	-.00007	.00	921	5	.55
HIT CAR-CUV-MINIVAN 3082+	-.0202	3.81	-.00138	.02	1,172	24	2.02
HIT TRUCK-BASED LTV < 4150	-.00483	.11	-.0397	8.62	543	3	.48
HIT TRUCK-BASED LTV 4150+	-.0237	3.25	-.0158	1.55	727	17	2.37
ALL OTHERS	-.0131	3.97	-.0102	2.45	2,552	33	1.31
VOLPE COEFFICIENT					9,499	65	.68

Test No. 6: Without Non-Significant Control Variables

	REGRESSION COEFFICIENTS					FATALITY INCREASE PER 100-POUND MASS REDUCTION	
	CURB WEIGHT		FOOTPRINT		FATALITIES AFTER ESC		
CRASH TYPE	COEFF	WALD CHI²	COEFF	WALD CHI²		N	%
			PICKUPS & TRUCK-BASED SUVs < 4,594				
1st-EVENT ROLLOVER	-.00782	2.03	-.0112	6.89	162	1	.78
HIT FIXED OBJECT	.0113	4.96	-.0186	24.16	451	-5	-1.13
HIT PEDESTRIAN/BIKE/MOTORCYCLE	-.00292	.56	.0055	6.16	676	2	.29
HIT HEAVY VEHICLE	-.0164	7.76	-.00823	5.22	287	5	1.64
HIT CAR-CUV-MINIVAN < 3082	-.00755	3.34	-.00285	1.54	530	-4	-.76
HIT CAR-CUV-MINIVAN 3082+	-.00833	3.58	-.00377	2.36	485	-4	-.83
HIT TRUCK-BASED LTV < 4150	-.00579	.49	-.00998	2.80	252	-1	-.58
HIT TRUCK-BASED LTV 4150+	-.0481	32.81	.0186	8.66	289	14	4.81
ALL OTHERS	-.00700	2.65	.00415	1.71	1,126	8	.70
VOLPE COEFFICIENT					**4,258**	**15**	**.35**
			PICKUPS & TRUCK-BASED SUVs ≥ 4,594				
1st-EVENT ROLLOVER	.00824	4.18	-.0112	6.89	345	-3	-.82
HIT FIXED OBJECT	-.00964	4.51	-.0186	24.16	802	8	.96
HIT PEDESTRIAN/BIKE/MOTORCYCLE	-.00438	1.82	.0055	6.16	1,516	-7	-.44
HIT HEAVY VEHICLE	-.00405	.53	-.00823	5.22	492	2	.41
HIT CAR-CUV-MINIVAN < 3082	.0140	19.55	-.00285	1.54	1,262	-18	-1.40
HIT CAR-CUV-MINIVAN 3082+	.0178	24.44	-.00377	2.36	1,155	-21	-1.78
HIT TRUCK-BASED LTV < 4150	.00992	3.42	-.00998	2.80	578	-6	-.99
HIT TRUCK-BASED LTV 4150+	-.00461	.66	.0186	8.66	490	2	.46
ALL OTHERS	.0030	1.11	.00415	1.71	2,262	-7	-.30
VOLPE COEFFICIENT					**8,902**	**-48**	**-.54**
			CUVs & MINIVANS				
1st-EVENT ROLLOVER	.0685	15.09	-.1154	39.12	100	-7	-6.85
HIT FIXED OBJECT	.0370	8.29	-.0887	42.78	373	-14	-3.70
HIT PEDESTRIAN/BIKE/MOTORCYCLE	.0221	4.87	-.00798	.31	812	-18	-2.21
HIT HEAVY VEHICLE	-.0304	3.44	-.0442	4.01	297	9	3.04
HIT CAR-CUV-MINIVAN < 3082	.00208	.04	-.00889	.72	503	-1	-.21
HIT CAR-CUV-MINIVAN 3082+	-.0110	.89	.00371	.11	569	6	1.10
HIT TRUCK-BASED LTV < 4150	-.0358	5.00	.0364	5.67	244	9	3.58
HIT TRUCK-BASED LTV 4150+	-.00853	.28	-.0276	2.67	294	-3	-.85
ALL OTHERS	.0020	.07	-.0181	5.08	1,380	-3	-.20
VOLPE COEFFICIENT					**4,571**	**-21**	**-.46**

Test No. 7: Including Muscle/Police/AWD cars and Full-Size Vans

REGRESSION COEFFICIENTS

CRASH TYPE	CURB WEIGHT		FOOTPRINT		FATALITIES AFTER ESC	FATALITY INCREASE PER 100-POUND MASS REDUCTION	
	COEFF	WALD CHI^2	COEFF	WALD CHI^2		N	%
			CARS < 3,106 POUNDS				
1st-EVENT ROLLOVER	.0162	1.83	-.0815	36.01	207	-3	-1.62
HIT FIXED OBJECT	.00322	.24	-.0458	41.11	813	-3	-.32
HIT PEDESTRIAN/BIKE/MOTORCYCLE	-.0245	8.86	-.0011	.01	871	21	2.45
HIT HEAVY VEHICLE	-.0250	4.84	-.0271	4.45	471	12	2.50
HIT CAR-CUV-MINIVAN < 3082	-.00961	.98	-.00016	.00	478	5	.96
HIT CAR-CUV-MINIVAN 3082+	-.00803	.77	-.00045	.01	674	5	.80
HIT TRUCK-BASED LTV < 4150	-.0172	2.22	-.0338	6.63	351	6	1.72
HIT TRUCK-BASED LTV 4150+	-.0628	33.91	-.0184	2.27	505	32	6.28
ALL OTHERS	-.0207	11.97	-.00896	1.86	1,530	32	2.07
VOLPE COEFFICIENT					5,901	107	1.81
			CARS ≥ 3,106 POUNDS				
1st-EVENT ROLLOVER	.0234	2.32	-.0815	36.01	247	-6	-2.34
HIT FIXED OBJECT	.0137	3.22	-.0458	41.11	1,263	-17	-1.37
HIT PEDESTRIAN/BIKE/MOTORCYCLE	.00092	.01	-.0011	.01	1,388	-1	-.09
HIT HEAVY VEHICLE	-.00647	.24	-.0271	4.45	687	4	.65
HIT CAR-CUV-MINIVAN < 3082	-.00051	.00	-.00016	.00	921	0	.05
HIT CAR-CUV-MINIVAN 3082+	-.0203	3.70	-.00045	.01	1,172	24	2.03
HIT TRUCK-BASED LTV < 4150	-.0100	.55	-.0338	6.63	543	5	1.00
HIT TRUCK-BASED LTV 4150+	-.0195	2.42	-.0184	2.27	727	14	1.95
ALL OTHERS	-.00898	1.72	-.00896	1.86	2,552	23	.90
VOLPE COEFFICIENT					9,499	47	.49

Test No. 7: Including Muscle/Police/AWD cars and Full-Size Vans

	REGRESSION COEFFICIENTS						
	CURB WEIGHT		FOOTPRINT		FATALITIES AFTER ESC	FATALITY INCREASE PER 100-POUND MASS REDUCTION	
CRASH TYPE	COEFF	WALD CHI^2	COEFF	WALD CHI^2		N	%
			PICKUPS & TRUCK-BASED SUVs < 4,594				
1st-EVENT ROLLOVER	-.00898	2.73	-.0116	7.72	162	1	.90
HIT FIXED OBJECT	.0119	5.35	-.0188	24.76	451	-5	-1.19
HIT PEDESTRIAN/BIKE/MOTORCYCLE	-.0124	4.93	.0137	11.61	676	8	1.24
HIT HEAVY VEHICLE	-.0204	6.26	-.00563	.90	287	6	2.04
HIT CAR-CUV-MINIVAN < 3082	.00291	.26	-.00109	.07	530	-2	-.29
HIT CAR-CUV-MINIVAN 3082+	.00995	2.71	-.0051	1.39	485	-5	-1.00
HIT TRUCK-BASED LTV < 4150	.0031	.15	-.00838	2.11	252	-1	-.31
HIT TRUCK-BASED LTV 4150+	-.0416	24.24	.0145	5.39	289	12	4.16
ALL OTHERS	-.00498	1.41	.00247	.66	1,126	6	.50
VOLPE COEFFICIENT					4,258	21	.49
			PICKUPS & TRUCK-BASED SUVs ≥ 4,594				
1st-EVENT ROLLOVER	.0229	26.89	-.0116	7.72	345	-8	-2.29
HIT FIXED OBJECT	-.00283	.41	-.0188	24.76	802	2	.28
HIT PEDESTRIAN/BIKE/MOTORCYCLE	.00373	.73	.0137	11.61	1,516	-6	-.37
HIT HEAVY VEHICLE	.00927	1.79	-.00563	.90	492	-5	-.93
HIT CAR-CUV-MINIVAN < 3082	.0120	7.47	-.00109	.07	1,262	-15	-1.20
HIT CAR-CUV-MINIVAN 3082+	.0184	15.76	-.0051	1.39	1,155	-21	-1.84
HIT TRUCK-BASED LTV < 4150	.0163	6.48	-.00838	2.11	578	-9	-1.63
HIT TRUCK-BASED LTV 4150+	-.00258	.13	.0145	5.39	490	1	.26
ALL OTHERS	.0033	.99	.00247	.66	2,262	-7	-.33
VOLPE COEFFICIENT					8,902	-68	-.76
			CUVs & MINIVANS				
1st-EVENT ROLLOVER	.0702	15.31	-.1159	22.66	100	-7	-7.02
HIT FIXED OBJECT	.0361	7.47	-.0767	18.19	373	-13	-3.61
HIT PEDESTRIAN/BIKE/MOTORCYCLE	.0157	2.19	-.00371	.07	812	-13	-1.57
HIT HEAVY VEHICLE	-.0194	1.10	-.0466	3.76	297	6	1.94
HIT CAR-CUV-MINIVAN < 3082	.0009	.01	.00787	.19	503	-0	-.09
HIT CAR-CUV-MINIVAN 3082+	-.0168	1.54	.0219	1.47	569	10	1.68
HIT TRUCK-BASED LTV < 4150	-.0382	3.94	.0405	2.48	244	9	3.82
HIT TRUCK-BASED LTV 4150+	.00927	.29	-.0380	2.64	294	-3	-.93
ALL OTHERS	.00396	.23	-.0272	5.98	1,380	-5	-.40
VOLPE COEFFICIENT					4,571	-17	-.37

Test No. 8: Controlling for Vehicle Manufacturer

	REGRESSION COEFFICIENTS					FATALITY INCREASE PER 100-POUND MASS REDUCTION	
	CURB WEIGHT		FOOTPRINT		FATALITIES AFTER ESC		
CRASH TYPE	COEFF	WALD CHI²	COEFF	WALD CHI²		N	%
			CARS < 3,106 POUNDS				
1st-EVENT ROLLOVER	.0182	1.41	-.0947	25.89	207	-4	-1.82
HIT FIXED OBJECT	-.00734	.80	-.0393	16.58	813	6	.73
HIT PEDESTRIAN/BIKE/MOTORCYCLE	-.0218	4.55	-.00406	.11	871	19	2.18
HIT HEAVY VEHICLE	-.0394	7.85	-.00422	.06	471	19	3.94
HIT CAR-CUV-MINIVAN < 3082	-.00943	.63	-.00363	.06	478	5	.94
HIT CAR-CUV-MINIVAN 3082+	-.00898	.64	-.00766	.33	674	6	.90
HIT TRUCK-BASED LTV < 4150	-.0179	1.56	-.0409	5.68	351	6	1.79
HIT TRUCK-BASED LTV 4150+	-.0646	23.12	-.0142	.77	505	33	6.46
ALL OTHERS	-.0157	4.57	-.0132	2.31	1,530	24	1.57
VOLPE COEFFICIENT					5,901	113	1.91
			CARS ≥ 3,106 POUNDS				
1st-EVENT ROLLOVER	.0271	2.23	-.0947	25.89	247	-7	-2.71
HIT FIXED OBJECT	.00936	1.02	-.0393	16.58	1,263	-12	-.94
HIT PEDESTRIAN/BIKE/MOTORCYCLE	-.00233	.04	-.00406	.11	1,388	3	.23
HIT HEAVY VEHICLE	-.0236	2.29	-.00422	.06	687	16	2.36
HIT CAR-CUV-MINIVAN < 3082	-.00134	.01	-.00363	.06	921	1	.13
HIT CAR-CUV-MINIVAN 3082+	-.0196	2.50	-.00766	.33	1,172	23	1.96
HIT TRUCK-BASED LTV < 4150	-.0166	1.07	-.0409	5.68	543	9	1.66
HIT TRUCK-BASED LTV 4150+	-.0337	5.02	-.0142	.77	727	24	3.37
ALL OTHERS	-.00493	.36	-.0132	2.31	2,552	13	.49
VOLPE COEFFICIENT					9,499	71	.75

Test No. 8: Controlling for Vehicle Manufacturer

	REGRESSION COEFFICIENTS					FATALITIES AFTER ESC	FATALITY INCREASE PER 100-POUND MASS REDUCTION	
	CURB WEIGHT		FOOTPRINT					
CRASH TYPE	COEFF	WALD CHI^2	COEFF	WALD CHI^2			N	%
			PICKUPS & TRUCK-BASED SUVs < 4,594					
1st-EVENT ROLLOVER	-.00023	.00	-.0154	12.24		162	0	.02
HIT FIXED OBJECT	.0171	10.09	-.0227	33.39		451	-8	-1.71
HIT PEDESTRIAN/BIKE/MOTORCYCLE	-.0147	5.99	.0152	11.85		676	10	1.47
HIT HEAVY VEHICLE	-.0137	2.41	-.00868	1.78		287	4	1.37
HIT CAR-CUV-MINIVAN < 3082	-.00236	.15	.00674	2.28		530	1	.24
HIT CAR-CUV-MINIVAN 3082+	-.00209	.11	.00276	.34		485	-1	-.21
HIT TRUCK-BASED LTV < 4150	.00031	.00	-.00523	.69		252	-0	-.03
HIT TRUCK-BASED LTV 4150+	-.0472	27.73	.0188	7.90		289	14	4.72
ALL OTHERS	-.00810	3.27	.00727	4.82		1,126	9	.81
VOLPE COEFFICIENT						4,258	29	.68
			PICKUPS & TRUCK-BASED SUVs ≥ 4,594					
1st-EVENT ROLLOVER	.00560	1.44	-.0154	12.24		345	-2	-.56
HIT FIXED OBJECT	-.00767	2.73	-.0227	33.39		802	6	.77
HIT PEDESTRIAN/BIKE/MOTORCYCLE	.00043	.01	.0152	11.85		1,516	-1	-.04
HIT HEAVY VEHICLE	-.00428	.30	-.00868	1.78		492	2	.43
HIT CAR-CUV-MINIVAN < 3082	.00639	1.85	.00674	2.28		1,262	-8	-.64
HIT CAR-CUV-MINIVAN 3082+	.0115	5.27	.00276	.34		1,155	-13	-1.15
HIT TRUCK-BASED LTV < 4150	.00427	.37	-.00523	.69		578	-2	-.43
HIT TRUCK-BASED LTV 4150+	-.00586	.61	.0188	7.90		490	3	.59
ALL OTHERS	-.00163	.21	.00727	4.82		2,262	4	.16
VOLPE COEFFICIENT						8,902	-12	-.13
			CUVs & MINIVANS					
1st-EVENT ROLLOVER	.0559	5.30	-.1117	13.56		100	-6	-5.59
HIT FIXED OBJECT	.0109	.40	-.0588	6.69		373	-4	-1.09
HIT PEDESTRIAN/BIKE/MOTORCYCLE	-.00409	.09	.0256	2.09		812	3	.41
HIT HEAVY VEHICLE	-.0229	.95	-.0381	1.61		297	7	2.29
HIT CAR-CUV-MINIVAN < 3082	-.0367	4.73	.0458	4.32		503	18	3.67
HIT CAR-CUV-MINIVAN 3082+	-.0445	6.96	.0465	4.42		569	25	4.45
HIT TRUCK-BASED LTV < 4150	-.0459	3.36	.0498	2.45		244	11	4.59
HIT TRUCK-BASED LTV 4150+	.0145	.41	-.0457	2.55		294	-4	-1.45
ALL OTHERS	-.0171	2.67	-.00763	.31		1,380	24	1.71
VOLPE COEFFICIENT						4,571	75	1.64

Test No. 9: Controlling for Vehicle Manufacturer and Nameplate

	REGRESSION COEFFICIENTS				FATALITIES AFTER ESC	FATALITY INCREASE PER 100-POUND MASS REDUCTION	
	CURB WEIGHT		FOOTPRINT				
CRASH TYPE	COEFF	WALD CHI^2	COEFF	WALD CHI^2		N	%
			CARS < 3,106 POUNDS				
1st-EVENT ROLLOVER	-.00353	.05	-.0581	8.43	207	1	.35
HIT FIXED OBJECT	-.0180	4.51	-.0176	2.89	813	15	1.80
HIT PEDESTRIAN/BIKE/MOTORCYCLE	-.0173	2.66	-.0102	.63	871	15	1.73
HIT HEAVY VEHICLE	-.0369	6.46	-.00145	.01	471	17	3.69
HIT CAR-CUV-MINIVAN < 3082	-.00996	.65	-.00086	.00	478	5	1.00
HIT CAR-CUV-MINIVAN 3082+	-.0121	1.08	.00018	.00	674	8	1.21
HIT TRUCK-BASED LTV < 4150	-.0129	.76	-.0466	6.66	351	5	1.29
HIT TRUCK-BASED LTV 4150+	-.0651	22.07	-.00845	.24	505	33	6.51
ALL OTHERS	-.0154	4.04	-.0111	1.44	1,530	24	1.54
VOLPE COEFFICIENT					5,901	122	2.07
			CARS ≥ 3,106 POUNDS				
1st-EVENT ROLLOVER	-.0111	.29	-.0581	8.43	247	3	1.11
HIT FIXED OBJECT	-.0233	4.89	-.0176	2.89	1,263	29	2.33
HIT PEDESTRIAN/BIKE/MOTORCYCLE	.00333	.07	-.0102	.63	1,388	-5	-.33
HIT HEAVY VEHICLE	-.0382	4.80	-.00145	.01	687	26	3.82
HIT CAR-CUV-MINIVAN < 3082	-.00589	.16	-.00086	.00	921	5	.59
HIT CAR-CUV-MINIVAN 3082+	-.0332	5.72	.00018	.00	1,172	39	3.32
HIT TRUCK-BASED LTV < 4150	-.0152	.72	-.0466	6.66	543	8	1.52
HIT TRUCK-BASED LTV 4150+	-.0499	8.70	-.00845	.24	727	36	4.99
ALL OTHERS	-.0118	1.65	-.0111	1.44	2,552	30	1.18
VOLPE COEFFICIENT					9,499	173	1.82

Test No. 9: Controlling for Vehicle Manufacturer and Nameplate

	REGRESSION COEFFICIENTS						FATALITY INCREASE PER 100-POUND MASS REDUCTION	
	CURB WEIGHT		FOOTPRINT			FATALITIES AFTER ESC		
CRASH TYPE	COEFF	WALD CHI^2	COEFF	WALD CHI^2			N	%
			PICKUPS & TRUCK-BASED SUVs < 4,594					
1st-EVENT ROLLOVER	-.00179	.10	-.0139	9.83		162	0	.18
HIT FIXED OBJECT	.0168	9.71	-.0225	32.25		451	-8	-1.68
HIT PEDESTRIAN/BIKE/MOTORCYCLE	.0160	7.02	.0163	13.39		676	11	1.60
HIT HEAVY VEHICLE	-.0126	2.01	-.00905	1.90		287	4	1.26
HIT CAR-CUV-MINIVAN < 3082	-.00019	.00	.00486	1.16		530	0	.02
HIT CAR-CUV-MINIVAN 3082+	.00343	.28	.00159	.11		485	-2	-.34
HIT TRUCK-BASED LTV < 4150	-.00028	.00	-.00453	.51		252	0	.03
HIT TRUCK-BASED LTV 4150+	-.0477	28.12	.0200	8.78		289	14	4.77
ALL OTHERS	-.00761	2.84	.00711	4.51		1,126	9	.76
VOLPE COEFFICIENT						4,258	28	.66
			PICKUPS & TRUCK-BASED SUVs ≥ 4,594					
1st-EVENT ROLLOVER	.00302	.40	-.0139	9.83		345	-1	-.30
HIT FIXED OBJECT	-.00797	2.79	-.0225	32.25		802	6	.80
HIT PEDESTRIAN/BIKE/MOTORCYCLE	-.00120	.06	.0163	13.39		1,516	2	.12
HIT HEAVY VEHICLE	-.00420	.27	-.00905	1.90		492	2	.42
HIT CAR-CUV-MINIVAN < 3082	.00976	4.13	.00486	1.16		1,262	-12	-.98
HIT CAR-CUV-MINIVAN 3082+	.0134	6.79	.00159	.11		1,155	-15	-1.34
HIT TRUCK-BASED LTV < 4150	-.00287	.16	-.00453	.51		578	-2	-.29
HIT TRUCK-BASED LTV 4150+	-.00873	1.28	.0200	8.78		490	4	.87
ALL OTHERS	-.0017	.22	.00711	4.51		2,262	4	.17
VOLPE COEFFICIENT						8,902	-12	-.13
			CUVs & MINIVANS					
1st-EVENT ROLLOVER	.0554	4.28	-.1146	12.89		100	-5	-5.44
HIT FIXED OBJECT	.00893	.23	-.0601	6.44		373	-3	-.89
HIT PEDESTRIAN/BIKE/MOTORCYCLE	-.00459	.09	.0243	1.69		812	4	.46
HIT HEAVY VEHICLE	-.0131	.28	-.0458	2.22		297	4	1.31
HIT CAR-CUV-MINIVAN < 3082	-.0223	1.51	.0301	1.72		503	11	2.23
HIT CAR-CUV-MINIVAN 3082+	-.0384	4.54	.0394	2.94		569	22	3.84
HIT TRUCK-BASED LTV < 4150	-.0500	3.50	.0516	2.45		244	12	5.00
HIT TRUCK-BASED LTV 4150+	.0271	1.23	-.0602	4.06		294	-8	-2.71
ALL OTHERS	-.0175	2.37	.00625	.19		1,380	24	1.75
VOLPE COEFFICIENT						4,571	60	1.31

Test No. 10: Limited to Drivers With BAC = 0

	REGRESSION COEFFICIENTS						
	CURB WEIGHT		FOOTPRINT		FATALITIES AFTER ESC	FATALITY INCREASE PER 100-POUND MASS REDUCTION	
CRASH TYPE	COEFF	WALD CHI²	COEFF	WALD CHI²		N	%
			CARS < 3,106 POUNDS				
1st-EVENT ROLLOVER	.0132	.55	-.0839	16.17	105	-1	-1.32
HIT FIXED OBJECT	-.00651	.44	-.0457	16.22	418	3	.65
HIT PEDESTRIAN/BIKE/MOTORCYCLE	-.0216	5.81	-.00708	.49	790	17	2.16
HIT HEAVY VEHICLE	-.0463	12.87	-.0141	.89	389	18	4.63
HIT CAR-CUV-MINIVAN < 3082	-.0175	2.47	-.00055	.00	394	7	1.75
HIT CAR-CUV-MINIVAN 3082+	-.00259	.06	-.0101	.72	569	1	.26
HIT TRUCK-BASED LTV < 4150	-.0195	2.28	-.0415	7.68	292	6	1.95
HIT TRUCK-BASED LTV 4150+	-.0666	30.59	-.0136	.95	437	29	6.66
ALL OTHERS	-.0226	11.85	-.00882	1.42	1,394	32	2.26
VOLPE COEFFICIENT					4,788	111	2.32
			CARS ≥ 3,106 POUNDS				
1st-EVENT ROLLOVER	.0182	.60	-.0839	16.17	107	-2	-1.82
HIT FIXED OBJECT	.00484	.17	-.0457	16.22	611	-3	-.48
HIT PEDESTRIAN/BIKE/MOTORCYCLE	-.00546	.27	-.00708	.49	1,240	7	.55
HIT HEAVY VEHICLE	-.0107	.51	-.0141	.89	572	6	1.07
HIT CAR-CUV-MINIVAN < 3082	-.00412	.10	-.00055	.00	767	3	.41
HIT CAR-CUV-MINIVAN 3082+	-.0178	2.15	-.0101	.72	1,012	18	1.78
HIT TRUCK-BASED LTV < 4150	-.00053	.00	-.0415	7.68	444	0	.05
HIT TRUCK-BASED LTV 4150+	-.0287	4.07	-.0136	.95	610	18	2.87
ALL OTHERS	-.0149	3.70	-.00882	1.42	2,287	34	1.49
VOLPE COEFFICIENT					7,651	81	1.06

Test No. 10: Limited to Drivers With BAC = 0

	REGRESSION COEFFICIENTS				FATALITIES AFTER ESC	FATALITY INCREASE PER 100-POUND MASS REDUCTION	
	CURB WEIGHT		FOOTPRINT				
CRASH TYPE	COEFF	WALD CHI^2	COEFF	WALD CHI^2		N	%
PICKUPS & TRUCK-BASED SUVs < 4,594							
1st-EVENT ROLLOVER	-.00793	1.16	-.0183	9.56	93	1	.79
HIT FIXED OBJECT	.00041	.00	-.0185	10.64	229	-0	-.04
HIT PEDESTRIAN/BIKE/MOTORCYCLE	-.0138	4.98	-.0148	10.63	603	8	1.38
HIT HEAVY VEHICLE	-.0251	7.11	-.00315	.21	236	6	2.51
HIT CAR-CUV-MINIVAN < 3082	-.00261	.18	-.00238	.27	452	-1	-.26
HIT CAR-CUV-MINIVAN 3082+	.0026	.15	-.00002	.00	400	-1	-.26
HIT TRUCK-BASED LTV < 4150	.0105	1.32	-.0112	2.94	207	-2	-1.05
HIT TRUCK-BASED LTV 4150+	-.0474	25.28	.0170	5.82	256	12	4.74
ALL OTHERS	-.00733	2.52	.0040	1.39	1,014	7	.73
VOLPE COEFFICIENT					3,491	30	.86
PICKUPS & TRUCK-BASED SUVs ≥ 4,594							
1st-EVENT ROLLOVER	.0327	28.20	-.0183	9.56	190	-6	-3.27
HIT FIXED OBJECT	-.00067	.01	-.0185	10.64	367	0	.07
HIT PEDESTRIAN/BIKE/MOTORCYCLE	.00174	.13	-.0148	10.63	1,341	-2	-.17
HIT HEAVY VEHICLE	-.00462	.29	-.00315	.21	423	2	.46
HIT CAR-CUV-MINIVAN < 3082	.00985	4.06	-.00238	.27	1,103	-11	-.99
HIT CAR-CUV-MINIVAN 3082+	.0149	8.13	-.00002	.00	994	-15	-1.49
HIT TRUCK-BASED LTV < 4150	.0162	4.97	-.0112	2.94	511	-8	-1.62
HIT TRUCK-BASED LTV 4150+	-.00613	.57	.0170	5.82	427	3	.61
ALL OTHERS	.00252	.47	.0040	1.39	2,030	-5	-.25
VOLPE COEFFICIENT					7,387	-43	-.58
CUVs & MINIVANS							
1st-EVENT ROLLOVER	.0998	21.69	-.1553	28.85	67	-7	-9.98
HIT FIXED OBJECT	.0395	5.99	.1017	21.92	259	-10	-3.95
HIT PEDESTRIAN/BIKE/MOTORCYCLE	.0203	3.38	-.00902	.35	756	-15	-2.03
HIT HEAVY VEHICLE	-.0369	3.41	-.0387	2.29	267	10	3.69
HIT CAR-CUV-MINIVAN < 3082	.00834	.37	.00526	.08	456	-4	-.83
HIT CAR-CUV-MINIVAN 3082+	-.0189	1.80	.0265	1.99	522	10	1.89
HIT TRUCK-BASED LTV < 4150	-.0492	5.79	.0488	3.28	219	11	4.92
HIT TRUCK-BASED LTV 4150+	.00986	.30	-.0385	2.52	268	-3	-.99
ALL OTHERS	.00025	.00	-.0256	4.98	1,291	-0	-.03
VOLPE COEFFICIENT					4,105	-8	-.19

Test No. 11: Limited to Good Drivers (BAC = 0, No Drugs, Valid License, Good Driving History)

| | REGRESSION COEFFICIENTS | | | | | | FATALITY INCREASE PER | |
| | CURB WEIGHT | | FOOTPRINT | | FATALITIES AFTER ESC | 100-POUND MASS REDUCTION | |
CRASH TYPE	COEFF	WALD CHI^2	COEFF	WALD CHI^2		N	%
				CARS < 3,106 POUNDS			
1st-EVENT ROLLOVER	-.00344	.03	-.0757	9.08	67	0	.34
HIT FIXED OBJECT	.0177	2.28	-.0445	10.71	291	5	1.77
HIT PEDESTRIAN/BIKE/MOTORCYCLE	.0262	6.72	-.00708	.39	624	16	2.62
HIT HEAVY VEHICLE	.0498	11.66	-.00981	.34	325	16	4.98
HIT CAR-CUV-MINIVAN < 3082	.0244	3.91	-.00105	.01	320	8	2.44
HIT CAR-CUV-MINIVAN 3082+	.0135	1.38	-.00132	.01	468	6	1.35
HIT TRUCK-BASED LTV < 4150	-.0188	1.71	-.0525	10.01	233	4	1.88
HIT TRUCK-BASED LTV 4150+	-.0693	27.08	-.00807	.28	362	25	6.93
ALL OTHERS	-.0295	16.90	-.00198	.06	1,177	35	2.95
VOLPE COEFFICIENT					3,868	116	3.00
				CARS ≥ 3,106 POUNDS			
1st-EVENT ROLLOVER	.0294	1.17	-.0757	9.08	75	-2	-2.94
HIT FIXED OBJECT	-.00664	.23	-.0445	10.71	452	3	.66
HIT PEDESTRIAN/BIKE/MOTORCYCLE	-.00737	.40	-.00708	.39	994	7	.74
HIT HEAVY VEHICLE	.0218	1.71	-.00981	.34	464	10	2.18
HIT CAR-CUV-MINIVAN < 3082	-.00286	.04	-.00105	.01	644	2	.29
HIT CAR-CUV-MINIVAN 3082+	.0275	4.27	-.00132	.01	857	24	2.75
HIT TRUCK-BASED LTV < 4150	.00988	.35	-.0525	10.01	367	-4	-.99
HIT TRUCK-BASED LTV 4150+	-.0372	5.76	-.00807	.28	511	19	3.72
ALL OTHERS	-.0223	7.08	-.00198	.06	1,982	44	2.23
VOLPE COEFFICIENT					6,345	103	1.62

Test No. 11: Limited to Good Drivers (BAC = 0, No Drugs, Valid License, Good Driving History)

	REGRESSION COEFFICIENTS				FATALITIES AFTER ESC	FATALITY INCREASE PER 100-POUND MASS REDUCTION	
	CURB WEIGHT		FOOTPRINT				
CRASH TYPE	COEFF	WALD CHI²	COEFF	WALD CHI²		N	%
	PICKUPS & TRUCK-BASED SUVs < 4,594						
1st-EVENT ROLLOVER	-.0183	4.49	-.00745	1.16	69	1	1.83
HIT FIXED OBJECT	.00046	.00	-.0166	6.35	170	-0	-.04
HIT PEDESTRIAN/BIKE/MOTORCYCLE	-.0116	2.83	.0146	8.22	467	5	1.16
HIT HEAVY VEHICLE	-.0330	9.97	-.00004	.00	190	6	3.30
HIT CAR-CUV-MINIVAN < 3082	.00543	.62	.00009	.00	383	-2	-.54
HIT CAR-CUV-MINIVAN 3082+	-.00582	.62	.00528	.96	338	2	.58
HIT TRUCK-BASED LTV < 4150	.0113	1.24	-.0134	3.40	142	-2	-1.13
HIT TRUCK-BASED LTV 4150+	-.0475	20.60	.0149	3.55	217	10	4.75
ALL OTHERS	-.0111	4.90	.00751	4.13	867	10	1.11
VOLPE COEFFICIENT					2,842	31	1.09
	PICKUPS & TRUCK-BASED SUVs ≥ 4,594						
1st-EVENT ROLLOVER	.0266	13.82	-.00745	1.16	143	-4	-2.66
HIT FIXED OBJECT	-.00789	.96	-.0166	6.35	265	2	.79
HIT PEDESTRIAN/BIKE/MOTORCYCLE	.00066	.01	.0146	8.22	1,080	-1	-.07
HIT HEAVY VEHICLE	-.00208	.05	-.00004	.00	342	1	.21
HIT CAR-CUV-MINIVAN < 3082	.0114	4.59	.00009	.00	915	-10	-1.14
HIT CAR-CUV-MINIVAN 3082+	.0122	4.51	.00528	.96	814	-10	-1.22
HIT TRUCK-BASED LTV < 4150	.0256	10.60	-.0134	3.40	424	-11	-2.56
HIT TRUCK-BASED LTV 4150+	-.0186	3.95	.0149	3.55	344	6	1.86
ALL OTHERS	-.00482	1.40	.00751	4.13	1,683	8	.48
VOLPE COEFFICIENT					6,010	-18	-.30
	CUVs & MINIVANS						
1st-EVENT ROLLOVER	.0982	17.05	-.1508	22.42	55	-5	-9.82
HIT FIXED OBJECT	.0239	1.62	-.0844	11.50	217	-5	-2.39
HIT PEDESTRIAN/BIKE/MOTORCYCLE	.0162	1.82	-.0010	.00	639	-10	-1.62
HIT HEAVY VEHICLE	-.0178	.69	-.0539	3.78	230	4	1.78
HIT CAR-CUV-MINIVAN < 3082	.00805	.30	.0047	.05	393	-3	-.81
HIT CAR-CUV-MINIVAN 3082+	-.0283	3.53	.0348	3.03	472	13	2.83
HIT TRUCK-BASED LTV < 4150	-.0295	1.86	.0157	.30	189	6	2.95
HIT TRUCK-BASED LTV 4150+	.00532	.08	-.0267	1.07	243	-1	-.53
ALL OTHERS	-.0021	.05	-.0204	2.86	1,155	2	.21
VOLPE COEFFICIENT					3,592	0	.00

DOT HS 811 665
August 2012

www.ingramcontent.com/pod-product-compliance
Lightning Source LLC
Chambersburg PA
CBHW080245180526
45167CB00006B/2420